Hazmatology
The Science of Hazardous Materials

Hazmatology: The Science of Hazardous Materials,
Five-Volume Set
9781138316072

Volume One - Chronicles of Incidents and Response
9781138316096

Volume Two - Standard of Care and Hazmat Planning
9781138316768

Volume Three - Applied Chemistry and Physics
9781138316522

Volume Four - Common Sense Emergency Response
9781138316782

Volume Five - Hazmat Team Spotlight
9781138316812

Standard of Care and Hazmat Planning

Robert A. Burke

CRC Press
Taylor & Francis Group
Boca Raton London New York

CRC Press is an imprint of the
Taylor & Francis Group, an **informa** business

CRC Press
Taylor & Francis Group
6000 Broken Sound Parkway NW, Suite 300
Boca Raton, FL 33487-2742

© 2021 by Taylor & Francis Group, LLC
CRC Press is an imprint of Taylor & Francis Group, an Informa business

No claim to original U.S. Government works

Printed on acid-free paper

International Standard Book Number-13: 978-1-138-31676-8 (Hardback)

**Visit the Taylor & Francis Web site at
http://www.taylorandfrancis.com**

**and the CRC Press Web site at
http://www.crcpress.com**

Printed in the United Kingdom
by Henry Ling Limited

Dedication

Volume Two

Max McRae

Max McRae, like Ron Gore, was not the one who thought of starting a hazmat team. However, he is the father of hazmat response in Houston. According to retired hazmat team member Bill Hand, "District Chief McRae was happy at Station 28. However, Chief McRae was the type of guy that would do what asked and do it to the best he could." McRae was the early force who brought the Houston hazmat team together and was there when the team was first placed in service. McRae became the first team coordinator, a position he retained until his retirement in August of 1994 after 40 years of service. Chief McRae has since passed away. In spite of their reluctance and apprehensions in the beginning, Chief McRae and Hand went on to do an outstanding job for the Houston hazmat team and had the respect and admiration of hazmat personnel throughout the United States and beyond.

Contents

Preface..xvii
Acknowledgements xix
Author... xxiii

Standard of Care and Hazmat Planning.. 1
  Hazards Defined 1
Dover, NJ, July 10, 1926, Picatinny Arsenal Munitions Explosion 1
Development of Hazmat Laws and Regulations.................................... 6
  How Legislative and Regulatory Process Works 6
  Legal Ramifications 10
Elements of the Hazardous Materials Standard of Care.......................... 11
  Legal Pit Falls... 11
  Definition of Liability 12
  Types of Liability... 12
  Types of Negligence... 13
  Sovereign Immunity... 14
  Regulatory Enforcement 15
    OSHA Enforcement.. 15
  Incidents That Caused Regulatory Change.. 16
    Fires 16
    Great Chicago Fire.. 17
    Baltimore City Fire 18
    Triangle Shirtwaist Fire, New York City... 19
    Our Lady of the Angels (OLA) Fire, Chicago, IL 20
    Hazardous Materials... 21
    Southwest Boulevard Fire, Kansas City, Kansas 22
    Marshalls Creek, PA, Explosion 24
  Bhopal, India, December 2–3, 1984, Release of Methyl Isocyanate
  "A Wake Up Call".. 25
    Background 25
    The Incident... 26
    Lessons Learned... 27
    Since 1984... 28

Institute West Virginia Methyl Isocyanate Gas Release........................ 29
    Definition of Standard of Care 30
Regulatory Basis for Hazardous Materials Standard of Care 30
    Hazardous Materials Laws and Regulations... 30
        Interstate Commerce Commission... 30
        Creation... 30
        Expansion of ICC Authority 31
    Federal Water Pollution Control Act Amendments of 1972................. 33
        Background: Why EPA Was Established 33
        Clean Water Act of 1970 and Amendments....................................... 34
        Summary of the Clean Water Act 33 U.S.C. §1251 et seq. (1972) 34
    Comprehensive Environmental Response, Compensation, and
    Liability Act (CERCLA) of 1980.. 35
        Superfund: CERCLA Overview 35
    The SARA aka Emergency Planning and Community
    Right-to-Know Act (EPCRA) 36
        EPCRA Overview... 37
        EPCRA Fact Sheet... 37
        What Are the Emergency Notification Requirements
        (Section 304)? 39
        What Are the Community Right-to-Know Requirements
        (Sections 311 and 312)? 40
        What Is the Toxics Release Inventory (Section 313)? 41
        Table 1: EPCRA Chemicals and Reporting Thresholds.................... 42
        What Else Does EPCRA Require?... 43
        Key Provisions of the Emergency Planning and Community
        Right-to-Know Act 44
    Hazardous Materials Transportation Uniform Safety Act of 1990
    (HMTUSA) 45
        National Oil and Hazardous Substances Pollution
        Contingency Plan (NCP) 46
        National Response Center.. 48
        National Response Team (NRT)... 49
        Supporting Regional Response Teams.. 51
        Regional Response Teams.. 51
        Responding to an Incident.. 53
        Federal Involvement.. 53
    EPA's Role in Emergency Response—Special Teams........................... 54
        Environmental Response Team.. 54
        Radiological Emergency Response Team 55
        Chemical, Biological, Radiological, and Nuclear Consequence
        Management Advisory Division... 55
        National Criminal Enforcement Response Team 55
        Summary of the Clean Air Act... 56

Clean Air Act...56
Clean Air Act of 1970 57
U.S. Patriot Act 58
Resource Conservation and Recovery Act (RCRA) Laws and
Regulations (RCRA)...62
What Is RCRA? 62
How Does RCRA Work?..62
Subtitle C—Hazardous Waste 63
Subtitle D—Nonhazardous Waste...63
RCRA Today 63
Federal Regulations 65
Standard of Care..66
Appendix E: Training Curriculum Guidance 68
Standards...70
Agencies...72
Department of Energy, DOT, Nuclear Regulatory
Commission and EPA 72
Government Resources..73
United States Chemical Safety and Hazard Investigation Board ... 73
Initial Operations 75
How to Report Incidents 76
Investigations Begin...76
Albert City, IA, April 8, 1990, Propane Explosion..............................77
United States Department of Transportation (DOT) 77
Regulations..77
Evolution of the Emergency Response Guidebook 78
Emergency Response Guidebook 78
Initial Response Actions..82
UN/DOT Classification System for Hazardous Materials...................88
Placard Requirements..89
Stenciled Commodities...90
DOT Chart 16 90
DOT Placard and Label System...90
Private Sector Resources and Regulations 94
Placard Hazard/Chemistry Chart..94
Fixed Facility Marking...96
Military Placard System 103
Additional Resources...104
Shipping Papers...104
Material Safety Data Sheets/Standard Operating Guidelines.......104
CHEMTREC...105
Do You Have an Emergency Involving Chemicals?..........................105
Emergency Call Center..106
CHEMTREC History...106

Assistance for Responders 110
Participation in Drills and Exercises 111
Planning for Hazardous Materials Incidents 111
Drexel Chemical Company Fire and Explosion.................................... 111
The Critical Importance and Implementation of ICS and Incident
Management 113
Introduction 113
Incident Command System (ICS)... 114
Incident Management Team (IMT).. 116
Command Staff.. 116
General Staff.. 116
Operations Section Chief 118
Planning Section Chief... 118
Logistics Section Chief 119
Finance/Administration Section Chief.. 119
Hazmat Incident Command... 120
Command.. 120
Organization and Function 120
Single Command... 120
Unified Command 121
ICS Structure.. 121
Branch 121
Division... 121
Group 121
Unit.. 121
Other Commands and Considerations 121
Area Command 121
Hazmat Incident Commander.. 121
First Arriving Unit Becomes the IC (OSHA 1910.120(q)(3)(i) 121
Safety Officer 123
Required (OSHA 1910.120(q)93)(vii) 123
Safety Officer Power to Terminate Unsafe Operations
(OSHA 1910.120(q)(3)(viii) 123
Transfer of Command... 124
Essential Elements of ICS.. 125
Common Terminology 125
Modular Organization.. 125
Management by Objectives... 125
Reliance on an IAP 126
Chain of Command 126
Unity of Command 126
Unified Command 126
Manageable Span of Control 127
Pre-designated Incident Locations and Facilities................................... 127

Resource Management 127
Integrated Communications 128
Transfer of Command 128
Accountability 128
Deployment 128
Command Post Location 128
Case Study Miamisburg, Ohio Phosphorus Incident 129
    Miamisburg, OH July 8, 1986, Derailment Phosphorus Fire 129
Safety 132
    Safety Officer 132
    Agency Liaisons 134
    Operations 134
    Research 134
    Decontamination (Decon) 134
        Forms of Decontamination 136
        Types of Decontamination 137
        Methods of Decontamination 138
    Contaminated Victims and Runoff Control 139
    Protected Personnel Following PPE Removal 139
        Gross 139
        Secondary 139
    Mass Decontamination Equipment 139
    Emergency Decontamination Equipment 139
    Decontamination Solutions 141
    Victim Transportation to Medical Facility 141
Case Study: Allegany County, PA Special Intervention Team 144
    Introduction 144
    False Assumptions 145
    Why Would We Want to Fix Something That Already Works? 145
    So How Does It Work? 146
    Removing Contamination 147
        Monitoring 149
        Hazmat EMS 150
    Requesting Additional Resources 150
        Request Assistance 150
    Federal On-Scene Coordinator 150
    Private Sector Assistance 151
    Establish Communications 151
    Joint Information Center (JIC) 151
    Public Information Officer (PIO) 151
Case Study: West, Texas Ammonium Nitrate Explosion 151
    Ammonium Nitrate Hazards 152
    Key Contributing Factors to Emergency Responders' Fatality 153
    Lack of Incident Command System 153

Lack of Established Incident Management System 154
Firefighter Training 154
Lessons Learned 156
Key Findings 156
Incident Management 157
Decision-Making Process 159
    Data 159
Recognition Primed Decision Model 159
Traditional Decision-Making Models 161
D.E.C.I.D.E. 161
    Detect HM Presence 161
    Estimate Likely Harm without Intervention 161
    Choose Response Objectives 161
    Identify Action Options 162
    Do Best Option 162
    Evaluate Progress 162
GEMBO 163
GEDAPER 165
    Gather Information 165
    Estimate Incident Course and Harm 165
    Determine Strategic Goals 165
    Assess Tactical Options and Resources 166
    Plan and Implement Actions 166
    Evaluate Operational Effectiveness 166
    Review Entire Process 167
Incident Priorities 167
Incident Levels 168
Scene Control Features 168
    Establish Perimeters 168
    Perimeter Distances 169
Isolation Zones 169
Access to Zones 170
Safe Refuge Ares 170
Public Protection Options 170
Case Study: Crete, NE February 19, 1969, Derailment
and Anhydrous Ammonia Release Shelter in Place
Effectiveness Substantiated 171
    Casualties 172
    Hatchetts 725 W. 13th Street 173
    Erdmans between 1005 and 1045 W. 13th Street 174
    Hoesche 1005 W. 13th Street 174
    Safranek 905 W. 13th Street 175
    Svarc 813 W. 13th Street 175
    Kovar 907 W. 13th Street 175

Svarc 915 Redwood Street 176
Crete Fire Department Responds 176
Crisis Management 180
Consequence Management.. 180
Emergency Operations Center (EOC) 180
Incident Generated Plans 181
Site Safety Plan 181
Plan of Action 181
Plan of Action Considers Existing Response Plans 182
Plan of Action 182
Terminating the Incident.. 183
After Action Analysis 184
After Action Report 184
After Action Follow-Up 184
Local Emergency Planning Committee (LEPC)............................... 184
LEPC Community Emergency Response Plan 184
Local Emergency Planning Committees 185
What Are the Required Elements of a Community
Emergency Response Plan?... 185
Commodity Flow Study 186
Conducting a Commodity Flow Study 186
Commodity Flow Study Highway Survey Checklist 191
Commodity Flow Study Railroad Survey Checklist 192
Site Specific Plans... 193
Elements to Include in Facility Response Plan 193
Site Specific Emergency Response Plan (ERP)................... 193
Procedures for Handling Emergency Incidents. 195
OSHA 1910.120 Emergency Response Plan 196
Hazardous Materials Containers.. 204
Highway Transportation Vehicles 204
Rail Road Transportation...211
Box Cars 212
Tank Cars.. 212
Water Transportation.. 219
Intermodal Containers 220
Fixed Facilities 224
Bulk Petroleum Storage... 225
Cone-Roof Tanks 225
Open Floating Roof Tanks 226
Closed Floating Roof Tanks.. 226
Horizontal Tanks.. 228
High-Pressure Tanks 230
Tube Banks 233
Vertical Cryogenic Tanks 233

Portable Containers..234

Incident History Is a Part of the Standard of Care 244
Historical Incidents 1800s 244
  Baltimore, MD, November 30, 1846, Powder Mill Explosion 244
  Wilmington, DE, August 3, 1855, Powder Mill Explosion...................245
  Washington, DC, June 18, 1864, Arsenal Explosion 245
  Cincinnati, OH, July 5, 1864, Train Explosion 246
  Coralville, IA, July 1875, Paper Mill Explosion 246
  McCainsville, NJ, July 2, 1886, Atlantic Dynamite Co. Explosion......247
  Rochester NY December 30, 1887, Naphtha Explosion 249
  Chicago IL, December 11, 1888, Oatmeal Mill Explosion 250
  Louisville, KY, June 30, 1890, Standard Oil Refinery Explosion.........251
  Kings Station, OH, July 16, 1890, Railroad Car Explosion...................252
  Blandford, VA, April 7, 1894, Fireworks Factory Explosion................252
  Elliotsville, WV, November 30, 1889, Powder Explosion.....................253
  Blue Island, IL, August 22, 1890, Standard Cartridge
  Factory Explosion 253
  Tarrytown, NY, May 22, 1891, Explosion and Train Wreck.................254
  Hartford, CT, May 21, 1892, Aetna Pyrotechnic Fire...........................255
  West Berkeley, CA, July 9, 1892, Powder Works Explosion.................256
  San Francisco, CA, May 22, 1895, Nitroglycerine
  Factory Explosion 257
  Murray, KY, February 24, 1897, Dynamite Explosion..........................257
  Cygnet, OH, September 8, 1897, Nitroglycerin Explosion 258
Historical Incidents 1900s 258
  Portland, OR, June 27, 1911, Union Oil Company Fire 258
  Waukegan, IL, November 27, 1912, Corn Products Plant Blast...........260
  Baltimore, MD, March 8, 1913 Harbor Dynamite Explosion...............261
  Buffalo, NY, June 25, 1913, Grain Elevator Explosion.........................263
  Cleveland, OK, March 9, 1916, Nitroglycerine Explosion 264
  Philadelphia, PA, September 8, 1917,
  Frankford Arsenal Explosions...265
  Jersey City, NJ August 17, 1915, NJ Oil Plant Explosion 266
  Ardmore, OK, September 27, 1915, Gasoline Explosion.......................266
  Syracuse, NY, July 5, 1918, Split Rock Explosion...............................267
  Vallejo, CA, November 8, 1918, Mare Island Navy Yard Explosion ...270
  Big Heart, OK, January 26, 1919, Nitroglycerine Explosion................270
  Cedar Rapids, IA, May 1919, Douglas Starch Co. Explosion.............271
  Whiting, IN, July 10, 1921, Standard Oil Explosion............................273
  Wellington, KS, June 7, 1922, Tank Car Fire.......................................274
  New York, NY, July 18, 1922, Chemical Explosion...............................274
  Altoona, KS, February 12, 1924, Nitroglycerine Explosion.................276
  Pittsburg, PA, August 20, 1924, Gasoline Explosion...........................276

Langtry, TX, March 5, 1925, Quarry Explosion 277
Clinton, IL February 21, 1925, Acid Explosion 277
Pensacola, FL, January 2, 1926, Newport Tar & Turpentine Co.
Explosion & Fire 278
Sparrows Point, MD, November 20, 1926, Oil Tanker Explosion 278
Alton, IL, June 15, 1926, Refinery Explosion 279
Turner, ID, January 19, 1927, Gas Explosion in Mormon Hall 280
San Pedro, CA, February 15, 1927, High School Classroom Blast 280
Chicago, IL, March 11, 1927, Chemical Company Explosion 281
Akron, OH, April 30, 1928, Benzene Explosion 281
Kokomo, IN, May 12, 1928, Explosion in Laundry 282
Elizabeth, NJ, February 19, 1930, Oil Refinery Explosion 283
Norfolk, VA, January 8, 1931, Oil Barge Explosion 284
Long Beach, CA, March 1, 1931, Gas Explosion 285
Kilgore, TX, April 18, 1931, Oil Tank Explosions 285
Marcus Hook, PA, February 5, 1932, Tanker Explosion 285
Berlin, NH, January 23, 1934, Oxygen Cylinder Explosion 287
Norman, OK, June 5, 1934, Dynamite Explosion 288
Asheville, NC, December 25, 1938, Fireworks Plant Blast 289
Owensboro, KY, November 12, 1939, Glenmore Distillery Fire 289
Camden, NJ, July 30, 1940, Camden Paint Factory Explosion 290
Toppenish, WA, November 29, 1940, Dynamite
Warehouse Explosion 290
Du Quoin, IL, February 15, 1941, Liquid Oxygen Plant Explosion 291
South Charleston, WV, Nov. 6, 1941, Plant Explosion 291
Browntown, WI, December 25, 1941, Silica Plant Explosion 292
Versailles, PA, May 2, 1942, Torpedo Plant Blast 292
Smithfield, NC, March 7, 1942, Ammunition Truck Explosion 294
Fontana, CA, December 25, 1942, Kaiser Steel Mills
Butane Explosion 295
Westport, CT, May 3, 1946, Trailer Truck Explosion 296
Greenville, SC, May 20, 1946, Propane Gas Explosion 297
Azusa, CA, August 21, 1946, Jet Plant Explosion 298
Bristol, TN, March 1, 1947, Filling Station Explosion 299
Waltham, MA, March 7, 1948, Plastics Plant Explosion 300
Sioux City, IA, December 15, 1949, Ammonia Plant Explosion 301
Crossville, IL, January 14, 1952, Propane Gas Explosion 303
Houston, TX, June 6, 1953, Fireworks Plant Explosion 303
Marianna, FL, August 14, 1953, Explosion and Fire Bottled Gas 305
Point Pleasant, WV, December 22, 1953, Gasoline Barge Explosion 307
Philadelphia, PA, October 10, 1954, Chemical Plant Explosion 307
Firemen Who Made the Supreme Sacrifice 308
Andale, KS, August 27, 1957, Farmers Coop Fire 308
Leonardo, NJ, May 23, 1958, Nike Missile Explosion 309

Orwigsburg, PA, June 3, 1959, Propane Truck Explosion 310
Speedway, IN, December 4, 1959, Air Products Plant Explosion.........311
Bayonne, NJ, December 29, 1960, Propane Explosion 312
Mitchell, IN, June 22, 1961, Lehigh Portland
Cement Co. Explosion.. 313
Berlin, NY, July 26, 1962, Propane Gas Explosion.................................314
Dunbar, PA, February 22, 1966, Fireworks Plant Explosion............... 315
Dunreith, IN, January 8, 1968, Train Wreck and Explosions 316
Blakely, GA, January 27, 1970, Butane Gas Explosion........................... 317
Brooklyn, NY, May 31, 1970, Liquid Oxygen Truck Explosion 318
Hollywood, FL, September 9, 1971, Dynamite Truck Explosion 318
East St. Louis, IL, January 23, 1972, Tank Car Explosion..................... 319
Sioux City, IA, April 30, 1974, Grain Elevator Explosion 319
Decatur, IL, July 19, 1974, Rail Yard Collision Fire and Explosion 321
Iowa City, IA, January 24, 1975, Propane Explosion............................ 324
Collinsville, IL, August 7, 1978, Propane Tank Car Explosion........... 325
Latham, KS, October 25, 1979, Propane Gas Explosion........................ 326
Raymond, NE, November 18, 1982, Propane Explosion 327
Rowesville, SC, May 25, 1983, Fireworks Truck Explosion................. 328
Santee, NE, August 28, 1983, Propane Gas Explosion......................... 329
Liquid Nitrogen Asphyxiation, Springer, OK, September 1998 330

Bibliography... 331
Index 335

Preface

Standards of care are nothing new to emergency response. Emergency medical personnel are bound by a standard of care based upon their level of training. It is enforced in local laws by the Emergency Medical Agency providing the care and by the medical director of that agency. Primary caregivers are emergency medical technicians (EMTs) and paramedics. Some additional responsibilities are given to EMTs if properly trained. However, both levels are allowed to only practice based upon the Standard of Care, which is largely based upon the amount of training and certification of the Emergency Medical Service (EMS) personnel. EMTs are limited on the level of life support they can give. Even if they know how to perform a procedure, such as cutting an opening in the throat to ventilate a choking victim, the Standard of Care prevents them from doing it. So, in some cases, a person may die if the proper level of life support is not available.

Emergency responders are also required to operate within a Standard of Care for response to hazardous materials and Weapons of Mass Destruction (WMD) incidents. It is based upon federal law, federal regulations, and consensus standards. It is also based upon lessons learned from previous incidents. Chemicals and compounds have been known to exist for centuries. Emergency responders have dealt with them for many years and often referred to them as chemicals or by their common chemical names, if known, such as gasoline, propane, ammonium nitrate, chlorine, and ammonia, to name a few. Initially, most of the knowledge gained about chemical responses came from experience, some good and some bad. Incidents have occurred over the years that resulted in the deaths and injuries of emergency response personnel operating at the scenes of chemical incidents—Texas City, TX (ammonium nitrate); Crescent City, IL (propane); Waverly, TN (liquefied petroleum gas); Kingman, AZ (propane); Kansas City, MO (gasoline); West, TX (ammonium nitrate); and many others. Once we have the appropriate level of hazmat training for the job we are being asked to do, we need to become familiar with the common hazardous materials in our communities that have historically killed emergency responders. Not every community will have all of them, but most will have some.

We should know as much as we can find out about historically dangerous common hazardous materials and their containers, and be familiar with the locations they are stored and used in our communities; perhaps we can prevent firefighter deaths and injuries resulting from these chemicals. Other chemicals will be present in many communities, most fairly common. Responders need to take the time to identify all of the major hazardous materials they may face in a local incident. Commodity flow studies can be conducted to determine which hazardous materials are shipped through our jurisdictions. Knowing all of them will be almost impossible. Statistically, the chances are relatively remote that any one community will experience a major hazmat incident from a transient hazardous material.

Planning is one of the most important tasks in all of hazmat response and community protection. Several types of plans will be discussed in Volume 2, including the site-specific plan, LEPC plan, your own hazmat response plan, and on-scene planning. Knowing your plans will help to more efficiently deal with any hazardous materials incident when and if one occurs. One of the most important aspects of the planning process is identifying what the hazardous materials in your community are, then how are they going to affect the community and emergency responders if they escape their containers, and finally, how are responders going to deal with the materials and bring the community back to normal again.

Acknowledgements

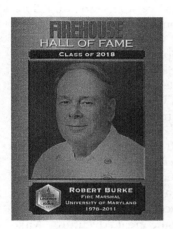

I thank the many fire departments and members across the United States and Canada that I have visited and became friends with during my visits to their departments over the years. I also thank the firefighters from classes I have attended as a student and taught for the National Fire Academy, Maryland Fire and Rescue Institute and Community College of Baltimore County since 1988. Learning is a two-way street, and I have learned much from the students as well. I thank the many friends I have met during the 40 plus years in the fire, EMS, hazardous materials and emergency management fields. There are those who I have not seen for a while; some are no longer with us, but once a friend, always a friend.

I express my thanks to *Firehouse Magazine* for allowing me to write stories about hazardous materials for 33 years and counting. During those years, I have had the pleasure of writing under every editor of the magazine including founder Dennis Smith who gave me the chance to be published for the first time. I also thank Firehouse editors, Janet Kimmerly, Barbara Dunleavy, Jeff Barrington, Harvey Eisner, Tim Sendelbach and Peter Mathews for their support over the years. When I read my first copy of *Firehouse Magazine* in the late 1970s, I was hooked. My dream was to someday go to Baltimore to attend a Firehouse Expo. Never did I dream I would not only attend an expo but teach at numerous expos, write for the magazine and in 2018 be inducted into the Firehouse Hall of Fame. To be placed in a fraternity with sixteen of the people who had an enormous impact on the fire service and who I looked up to my entire career was very humbling.

Several people have been my mentors and have impacted my life and career. When I worked with the State Fire Marshall of Nebraska, Wally Barnett allowed me to accomplish things in the State Fire Marshal's Office

Brent Boydston, Chief
Bentonville, AR Fire
Department.

that I otherwise would not have. Because of his ability to let his employees reach their potential, I was able to write for *Firehouse Magazine*, become a contributing editor, teach for the National Fire Academy and other things too numerous to mention. He was proud when I gave him a copy of my first book. I owe much of my success in the fire service to the opportunities Wally gave me. Jan Kuczma and Chris Waters at the National Fire Academy have been mentors to me over the years. Ron Gore, retired Captain from the Jacksonville, FL Fire Department and Owner of Safety Systems, has had a large impact on my life and career. The Jacksonville Hazmat Team was the first emergency services Hazmat Team in the United States. Ron Gore is the God Father of Hazmat response in the United States.

Former student of mine and current Chief of the Bentonville, AR Fire Department Brent Boydston has been a great friend to me and my family over the years. Rudy Rinas, Gene Ryan and John Eversole of the Chicago Fire Department have been fellow classmates and students. Mike Roeshman and Bill Doty of the Philadelphia Fire Department are both former students of mine and are now retired as Hazmat Chief Officers have remained friends. I used to ride with Bill and together we had some great adventures. Mike showed me Philadelphia historical areas, like the spot where Ben Franklin flew his kite and his post office, which is so obscure today in downtown Philadelphia. I also stood on the spot where Rocky stood at the top of the

Mike Roeshman Retired
Hazmat Chief Philadelphia
Fire Department.

steps in the movie. These adventures enjoyed in Philadelphia would not have happened without Bill and Mike.

Just outside of Philadelphia in Delaware County, Tom Micozzie, Hazmat Coordinator for Delaware County, was also a former student and a great friend. We had many adventures together, and I will never forget his introduction to me of the Galati at Rita's Italian Ice! Rita's Italian Ice was started by a retired Philadelphia firefighter and not long ago one opened up in Lincoln, NE.

Thanks to Richmond In Fire Chief Jerry Purcell, who I met during a visit to Richmond to do a Firehouse story on their 1968 explosion in downtown. As a result of

William, "Bill" Doty retired Hazmat Chief Philadelphia Fire Department.

the Richmond story being published I was able to locate and become friends with blast survivor Jack Bales. More recently I visited to do another story on their hazmat team and propane training. Thanks to new friend Ron Huffman who traveled to Richmond to conduct the propane training utilizing water injection to control liquid propane leaks. The article appeared in the September 2019 *Firehouse Magazine*.

Thanks to Tod Allen, Fire Chief in Crete Nebraska who I met when I was researching a train derailment in Crete for another friend Kent Anderson. We have become good friends. Tod is the apparatus operator on Truck one at Station 1 for the Lincoln Nebraska Fire Department. He invited me to come and ride with him, and many adventures later I still go there on a regular basis. I thank all of my friends past and present on "B" Shift at Station 1 for making me feel at home and showing me a good time whenever I am there. Thanks to friend Captain Mark Majors for sharing his experiences with Nebraska Task Force 1 Urban Search and Rescue Team (USAR) and Captain Francisco Martinez assigned to Station 14 Lincoln Hazmat has been a great help with gathering information and taking photos for me. Finally, I thank Chief Michael Despain and assistant Chief Patrick Borer for their friendship and hospitality while visiting the Lincoln Fire Department on many occasions. This is only the short list—I would have to write a separate book to thank all of you I have met and for the impact you have had on my

Chief Jerry Purcell Richmond, IN Fire Department.

life over the past 40+ years. You know who you are; I appreciate your friendship and assistance and consider your selves thanked again.

During my year-long book writing adventure that led to *Hazmatology: The Science of Hazardous Materials*, I met and spoke to many people and made new friends. I thank my cousin Dustin Schroeder, Senior Captain at Houston Station 68, the firefighters and others I met while in Houston. I also thank Kevin Okonski, Hazmat at Houston Station 22; Ludwig Benner, former NTSB Investigator and developer of several incident management models; Bill Hand, Houston; Richard Arwood; Charles Smith,

Memphis; Kevin Saunders, Motivator; Chief Jeff Miller, Butte, MT; and all of the Nebraska Regional Team leaders and members.

I express my thanks to my cousin Jeanene and her husband Randy for coming all the way from Montana to be with me at the Firehouse Hall of Fame induction. I am also grateful to Brent Boydston, James Rey Milwaukee, Wilbur Hueser and Saskatoon in Canada for the hospitality and tour, and Captain Oscar Robles, Imperial, CA. The list just goes on and on, and there is not room here to name everyone, but the rest of you know who you are and I want you to know how much your assistance is appreciated. You are all considered friends, and I hope we will talk and or meet again. Finally, thanks to librarians and historians across the country for your assistance in research, thanks for the memories!

Robert Burke

Author

 **Robert A. Burke** was born in Beatrice and grew up in Lincoln, NE; graduated from high school in Dundee, IL; and earned an AA in fire protection technology from Catonsville Community College, Baltimore County, MD (now Community College of Baltimore County) and a BS in fire administration from the University of Maryland. He has also completed his graduation in public administration at the University of Baltimore. Mr Burke has attended numerous classes at the National Fire Academy in Emmitsburg, MD, and additional classes on firefighting, hazardous materials, and weapons of mass destruction at Oklahoma State University; Maryland Fire and Rescue Institute; Texas A&M University, College Station, TX; the Center for Domestic Preparedness in Anniston, AL; and others.

Mr. Burke has over 40 years' experience in the emergency services as a career and volunteer firefighter. He has served as a Lieutenant for the Anne Arundel County, Maryland Fire Department; an assistant fire chief for the Verdigris Fire Protection District in Claremore, OK; Deputy State Fire Marshal in the State of Nebraska; a private fire protection and hazardous materials consultant; and an exercise and training officer for the Chemical Stockpile Emergency Preparedness Program (CSEPP) for the Maryland Emergency Management Agency; and retired as the Fire Marshal for the University of Maryland. He has served on several volunteer fire companies, including West Dundee, IL; Carpentersville, IL; Sierra Volunteer Fire Department, Chaves County, NM; Ord, NE; and Earleigh Heights Volunteer Fire Company in Severna Park, MD, which is a part of the Anne Arundel County, MD, Fire Department.

Mr. Burke has been a Certified Hazardous Materials Specialist (CFPS) by the National Fire Protection Association (NFPA) and certified by the National Board on Fire Service Professional Qualifications as a Fire

Instructor III, Hazardous Materials Incident Commander, Fire Inspector III, and Plans Examiner II. He served on the NFPA technical committee for NFPA 45 Fire Protection for Laboratories Using Chemicals for 10 years. He has been qualified as an expert witness for arson trials as well.

Mr. Burke retired as an adjunct instructor at the National Fire Academy in Emmitsburg, MD, in April 2018 after 30 years. He taught hazardous materials, weapons of mass destruction, and fire protection curriculums. He taught at his Alma Mater Community College of Baltimore County, Catonsville Campus, and Howard Community College in Maryland. He has had articles published in various fire service trade magazines for the past 31 years. Mr. Burke is currently a contributing editor for *Firehouse Magazine*, with a bimonthly column titled "Hazmat Studies," and he has had numerous articles published in *Firehouse, Fire Chief, Fire Engineering,* and *Nebraska Smoke Eater* magazines. He was inducted into the Firehouse Hall of Fame in October 2018 in Nashville, TN. Mr Burke has also been recognized as a subject matter specialist for hazardous materials and been interviewed by newspapers, radio, and television about incidents that have occurred in local communities including Fox Television in New York City live during a tank farm fire on Staten Island.

Mr. Burke has been a presenter at Firehouse Expo in Baltimore, MD, and Nashville, TN, numerous times, most recently in 2017. He gave a presentation at the EPA Region III SERC/LEPC Conference in Norfolk, VA, in November 1994, and a presentation at the 1996 Environmental and Industrial Fire Safety Seminar, Baltimore, MD, on DOT ERG. He was a speaker at the 1996 International Hazardous Materials Spills Conference, on June 26, 1996, in New Orleans, LA; a speaker at the 5th Annual 1996 Environmental and Industrial Fire Safety Seminar in Baltimore, MD, sponsored by Baltimore City Fire Department; and at Local Emergency Planning Committee (LEPC), an instructor for hazmat Chemistry, August 1999, at Hazmat Expo 2000 in Las Vegas, NV. He gave a keynote presentation at the Western Canadian Hazardous Materials Symposium Saskatoon, Saskatchewan, Canada, in 2008.

Mr. Burke has developed several CD-ROM-based training programs, including the Emergency Response Guide Book, Hazardous Materials and Terrorism Awareness for Dispatchers and 911 Operators, Hazardous Materials and Terrorism Awareness for Law Enforcement, Chemistry of Hazardous Materials Course, Chemistry of Hazardous Materials Refresher, Understanding Ethanol, Understanding Liquefied Petroleum Gases, Understanding Cryogenic Liquids, Understanding Chlorine, and Understanding Anhydrous Ammonia. He has also developed the "Burke Placard Hazard Chart." He has published seven additional books titled *Hazardous Materials Chemistry for Emergency Responders* (first, second, and third editions); *Counterterrorism for Emergency Responders* (first, second and third editions); and *Fire Protection: Systems and Response and Hazmat Teams across America.*

Currently, Mr. Burke serves on the Homestead LEPC in Southeast Nebraska. He manages a Hazardous Materials section at the Nebraska Firefighters Museum and periodically rides with friends on "B" shift at Station 1, Lincoln Fire Department. He can be reached via email at robert. burke@windstream.net, on Facebook at https://www.facebook.com/ RobertAb8731, and through his website: www.hazardousmaterialspage. com.

Volume Two

Standard of Care and Hazmat Planning

Hazards Defined

Statistics and details for fixed site and rail transportation incidents involving chemicals go back to the mid-1800s. Hundreds of people died in rail accidents in the 1800s and early 1900s. None of them involved chemicals; they all died in derailments of passenger trains. Not until 1959 was the first recorded derailment involving the release of hazardous materials, where liquefied petroleum gas (LPG) (one of the common chemicals) escaped from its container and killed 23 civilians in Meldrim, GA. This incident is also known as the "Meldrim trestle disaster." Since that time, the Interstate Commerce Commission and its predecessor, the National Transportation Safety Board (NTSB), have reported just 44 train derailments where chemicals escaped their containers. Not much data is available outside of searching news clippings on fixed facility chemical releases. Newspaper research for *Hazmatology: The Science of Hazardous Materials* has revealed some interesting statistics. My first discovery of a chemical incident that involved an explosion occurred in 1841 in Syracuse, NY. Twenty-five kegs of black powder exploded, killing over 30 civilians and injuring over 50. Explosives were typical sources of deaths among civilians and firefighters through the 1800s well into the 1900s. Explosive incidents increased in times of war as ammunition plants geared up production to meet the war needs. These incidents were noticed around the Civil War, World War I and World War II.

Dover, NJ, July 10, 1926, Picatinny Arsenal Munitions Explosion

On July 10, 1926, an explosion occurred at DNASD that killed nearly two dozen people, shaking the U.S. military to its core (Figure 2.1). The cause was not mechanical failure, human error, or sabotage; but it was Mother Nature. A single lightning strike during a thunderstorm was the likely source. Shortly after 5:00 in the evening, the thunderstorm produced a bolt of lightning that struck the storage depot at Picatinny Arsenal. More than 600,000 tons of explosives stored inside the depot detonated, resulting in

Figure 2.1 On July 10, 1926, an explosion occurred at DNASD that killed nearly two dozen people, shaking the U.S. military to its core.

one of the most catastrophic explosions in the United States. The blast completely destroyed nearly 200 buildings in a 1/2-mile radius, resulting in $47 million in damages (more than $631 million today), 21 deaths and dozens more injuries. The explosion was so powerful that people reported finding debris nearly 22 miles away.

The incident at the Picatinny Arsenal prompted the U.S. government to get serious about explosives safety. Shortly after the explosion, Congress created the Department of Defense Explosives Safety Board, a board that exists to "provide oversight of the development, manufacture, testing, maintenance, demilitarization, handling, transportation and storage of "explosives" within the military." The board exists to this day.

Late in the 1860s, oil was discovered in Pennsylvania which brought a new source of death and destruction. From then on, it was mostly oil and explosives that caused death and injury. In those days, there was little if any regulation of chemicals or zoning requirements. Explosives use, storage and manufacturer could be found in residential neighborhoods and main street business districts. Televisions and radios did not exist and traveling performers were scarce. Citizens flocked to fires to socialize and be entertained. Unfortunately, many explosions happened after the fire department and crowds were at the scene of a fire. Explosions and chemical incidents involved common materials in place at the time. History seems to have repeated itself over the years since 1841.

Medical care is likely another reason why so many civilians and firefighters died in fires and explosions from 1841 until the mid-1900s. Not every community had access to a hospital. Even when they did, hospitals in those days did not have mass casualty plans. Doctors were in short supply in many communities, and on-scene emergency medical care was still over a hundred years away. Undertakers had the closest thing to an ambulance, and that continued into the 1960s in some areas of the United States.

Author's Note: Just out of high school, one of my jobs was working for a furniture store doing odd jobs and delivering furniture. For reasons unknown to me, it was not uncommon for undertakers to also own a furniture store. This furniture store in Davenport, Nebraska, was owned by the undertaker. When I was hired, he found out I had been a volunteer firefighter for a couple of years. After that it was my job to drive the ambulance when it was requested. The Cadillac hearse was turned into an ambulance by simply installing a light into a custom bracket on top of the vehicle. The siren was hidden under the hood, and switches for each were installed on the dash board. My only training was Boy Scout First Aid, and my only proof my merit badge. Emergency medical technicians wouldn't be invented for another 5 years in 1973 when I first became an Emergency Medical Technician (EMT). So, for now it was load and go.

Local, state and federal laws and regulations were slow coming, which eventually helped curb the numbers of explosions involving explosives in manufacture, transportation and storage. Zoning took even longer. Following the evolution of unions, in addition to pay issues, they also were concerned about worker safety. Although their only tool to make progress in safety issues was work stoppages, which often became violent in nature, companies would hire private security firms like Pinkerton's for strike busting purposes.

After the discovery of oil, it wasn't long before the invention of vehicles powered by engines utilizing petroleum-based fuels, including the motorization of many fire departments (Figure 2.2). Now in addition to railroads, another form of transportation was evolving. Prior to hazardous materials regulations, there were a number of transportation accidents on the road and the rail, and fixed facilities that killed firefighters into the 1980s. Railroads began policing themselves and developing safety devices to make railroad transportation of hazardous materials safer. Creation of the U.S. Department of Transportation (DOT) resulted in tools for emergency responders to help identify containers and hazards of chemicals.

Following the chemical release in Bhopal, India, mandated training and response regulations in the United States made hazardous materials response safer and more efficient. Since the formation of the Chemical Safety Board (CSB) in 1998, there have been 133 major incidents involving escaping or exploding chemicals at fixed facilities. Highway transportation

Figure 2.2 After the discovery of oil, it wasn't long before the invention of vehicles powered by engines utilizing petroleum-based fuels, including the motorization of many fire departments. (Courtesy of Beatrice NE Fire Department.)

Figure 2.3 Highway transportation incidents occur more frequently than rail or fixed facility incidents, and some very serious ones have killed firefighters. (Courtesy of Marshalls Creek Fire Department.)

incidents occur more frequently than rail or fixed facility incidents, and some very serious ones have killed firefighters (Figure 2.3). However, the vast majority of them are handled successfully by firefighters and hazardous materials responders.

The modern-day coinage of the term "hazardous material" occurred in the mid-1970s, when the DOT established a definition of hazardous

Figure 2.4 The modern-day coinage of the term "hazardous material" occurred in the mid-1970s, when the DOT established a definition of hazardous material. (From DOT.)

material (Figure 2.4). The term "dangerous goods," which means the same thing, is used outside the United States. DOT began the first major regulation of hazardous materials in transportation, including a hazard-class and placard and label system for identifying hazardous materials. As other federal agencies began developing regulations dealing with hazardous materials storage and use, different names were also created. The U.S. Occupational Safety and Health Administration (OSHA) and the U.S. Environmental Protection Agency (EPA) both refer to hazardous materials as "hazardous substances." The EPA also regulates chemicals that no longer have a commercial value.

When chemicals are no longer useful for their intended purpose, they become hazardous waste. Hazardous waste is regulated in the workplaces where it is generated, during transportation to a disposal site and when it is disposed of. For example, gasoline (Figure 2.5), when transported, is a hazardous material regulated by the DOT. When a tanker offloads gasoline into an underground storage tank at a gasoline station, it becomes a hazardous substance regulated by the EPA and OSHA. If any gasoline is spilled on the ground during the offloading, it would become hazardous waste, regulated by OSHA, EPA and DOT. There are different names for the same gasoline, depending on whether it was transported, in fixed storage or spilled. For the purposes of this column, we will use the term "hazardous material" interchangeably with all other agency terminologies.

After the Bhopal incident, the U.S. Congress passed the Emergency Planning and Community Right-to-Know Act of 1986 (EPCRA), also known as the Superfund Amendments and Reauthorization Act (SARA). Congress was concerned that such an incident could happen here.

Figure 2.5 Gasoline, when transported, is a hazardous material regulated by the DOT.

Additionally, Congress was also concerned about the level of preparedness and training available to deal with an incident of the magnitude of Bhopal. With the passage of this important legislation, the federal government for the first time mandated levels of training and competency for emergency responders to hazardous materials releases.

Development of Hazmat Laws and Regulations

How Legislative and Regulatory Process Works

Laws are enacted by a legislative body, such as Congress, state legislatures, county governing boards and local city councils and boards. These legislative bodies have no means to enforce the laws they pass. If laws must be enforced, that task is passed on to an enforcement or regulatory agency that forms procedures or regulations to implement the law. On the federal level, laws that concern hazardous materials are generally passed on to the U.S. Occupational Safety and Health Administration (OSHA), Environmental Protection Agency (EPA) or Department of Transportation (DOT).

When the Emergency Planning and Community Right-to-Know Act of 1986 (EPCRA) was passed by Congress, OSHA and EPA were tasked with developing regulations to implement the requirements of the act. OSHA 1910:120 and EPA 40 CFR 311 are identical regulations dictated by federal laws that apply to emergency responders that may respond to hazmat incidents.

In simple terms, the regulations determine what emergency respond-ers are allowed to do and what they are not. Some states have a delegated authority to enforce OSHA regulations. In those states, hazmat regula-tions are enforced by the state OSHA. In states that do not have a state OSHA, the EPA regulations are enforced. So, whether responders are in an OSHA state or a non-OSHA state, they are covered by the federal regu-lations for hazmat responses.

National Fire Protection Association (NFPA) standards, while they are not laws, are consensus standards developed by committees that deter-mine what is appropriate for each level of hazmat response and what is not. A jurisdiction may implement NFPA standards, which makes them required in that jurisdiction, much like a regulation. While NFPA stan-dards are not laws, they are the recognized way of addressing issues that face the fire service and other organizations today, including hazmat responses. Hazmat first responders may include fire, police, EMS, public works industrial personnel and other public and private workers.

OSHA and EPA regulations governing training requirements for hazmat responses establish five levels of competency for hazardous materials responders. According to OSHA 1810.120, "competent" means possessing the skills, knowledge, experience and judgment to perform assigned tasks or activities satisfactorily as determined by the employer. All responders must be trained specifically for the level at which they are expected to perform by their employers. Employers are charged with the responsibility of determining the level of response and what training is required for that level, and developing standard operating procedures (SOPs) or standard operating guidelines (SOGs) for hazmat responses in their jurisdictions.

There are five levels of hazardous materials responder training:

- First responder awareness
- First responder operations
- Technician
- Specialist
- Incident commander

These five levels are also used in the NFPA 472 Standard for Competence of Responders to Hazardous Materials/Weapons of Mass Destruction Incidents. The NFPA standard provides much more detail than does OSHA 1910.120. Additional competencies for emergency medical person-nel are outlined in NFPA 473. Each level of response has associated with it certain competencies and limitations placed upon emergency responders for their safety. Along with the legislation and standards mentioned above comes an implied "Standard of Care" associated with responses to haz-ardous materials incidents. A Standard of Care is the level of competence anticipated or mandated during the performance of a service or duty. A Standard of Care is not static but constantly changing, influenced by laws, regulations, consensus standards, knowledge and experience. Standards of Care are not new to emergency response personnel.

Hazmat response personnel have limitations on what functions they can perform at the scene of a hazardous materials incident. Limitations are based upon the level of knowledge, experience, training and the availability of personnel protective equipment (PPE) and supplies. Awareness, operations, technician, specialist and incident command personnel may also have limitations, based upon the jurisdiction with which they belong. Recognition of the existence of hazardous materials is the single most important task any emergency responder can do upon arrival at an incident scene. Generally, hazardous materials scenes are divided into zones for the safety of personnel. These may include "cold," "warm" and "hot" zones (Figure 2.6). The hot zone is where the hazardous materials are located, and the greatest danger exists for response personnel. The "warm" zone is where decontamination takes place. The "cold" zone is everywhere else and should not present an immediate danger to personnel.

According to OSHA, awareness-level personnel are those who, in the course of their normal duties, could encounter an emergency involving hazardous materials and who are expected to recognize the presence of the hazardous materials, protect themselves, call for trained personnel and secure the area. An example would be department of roads or public works, law enforcement and utilities personnel. When emergency response organizations respond to a hazmat incident, OSHA has said in a formal interpretation (July 25, 2007) that those responders should be operations-level personnel who first arrive at the scene of the incident. Operations-level personnel are those who respond to a hazmat incident for the purpose of protecting nearby persons, the environment or property from the effects of the release. Operations-level personnel are required to

Isolation Zones

- Hot Zone – **Where hazardous materials are located**
- Warm Zone – **Where decontamination takes place**
- Cold Zone – **Everywhere else**

Figure 2.6 Generally, hazardous materials scenes are divided into zones for the safety of personnel. These may include "cold," "warm" and "hot" zones.

have awareness training as well as an additional 8 h of operations-level training. Awareness- and operations-level personnel do not enter the hot zone.

Technician-level responders are members of organized hazmat response teams (Figure 2.7). They may enter the hot zone and work in close proximity to hazardous materials if they have the proper chemical protective clothing, respiratory protection, mitigation equipment and training. Specialist personnel are those who are trained to the technician level and have additional training in an area of expertise such as rail cars or particular chemicals. The incident commander level requires training to a minimum of the awareness and operations levels (minimum 24 hours) and competencies outlined in OSHA 1910.120 and NFPA 472 for that position (Figure 2.8).

NFPA 472 identifies the incident commander as the person responsible for all incident activities, including the development of strategies and tactics and the ordering and release of resources. OSHA philosophy on dealing with hazardous materials is actually quite simple—they want employers to train and equip their employees for the jobs they are called upon to do. Some response organizations, because of reduced staffing levels, use operations-level personnel to conduct decontamination.

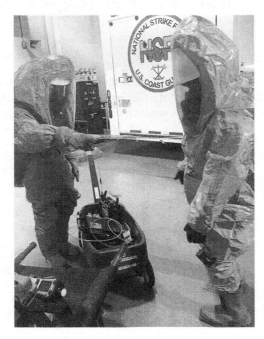

Figure 2.7 Technician-level responders are members of organized hazmat response teams. (From U.S. Coast Guard Atlantic Strike Team.)

Figure 2.8 The incident commander level requires training to a minimum of the awareness and operations levels (minimum 24 h).

These personnel are not technicians but rather trained and equipped operations-level personnel. This is an acceptable practice because operations-level personnel are trained and equipped to do decontamination (OSHA, NFPA).

Legal Ramifications

In today's litigious society, even emergency responders and response organizations can be sued for not following a Standard of Care (Figure 2.9). Several legal terms should be known by responders and response organizations. The first term is liability, which is defined as owing a responsibility

Figure 2.9 In today's litigious society, even emergency responders and organizations can be sued for not following a Standard of Care.

or duty to act. All emergency responders have a responsibility or duty to act when called to the scene of an incident involving hazardous materials. Those responsibilities should be outlined in the organization's SOPs. The only way to eliminate that liability would be to quit being an emergency responder.

If response personnel operate outside the organization's SOPs or the hazardous materials Standard of Care, they may be determined to be negligent in the performance of their duties. Negligence is the performance outside of the accepted Standard of Care. Negligence can be the fault of the responder, officer, organization or employer. It is possible to be negligent without knowing it. However, I am sure you have heard the saying, "Ignorance of the law is no excuse." Gross negligence occurs when a person, officer, organization or employer willfully operates outside of the Standard of Care, for example, if the organization/employer does not provide the required level of training for employees. More important, response personnel may be injured or killed if not following local SOPs or the hazardous materials Standard of Care. Hazmat response organizations, officers and personnel should be aware of and implement the requirements of OSHA 1910.120 and NFPA standards 472 and 473 in their operations.

> ***Author's Note:*** *In my opinion, it is pretty simple. I have always believed that the basic premise of the Hazardous Materials Standard of Care is that responders must be trained and equipped to do the job they are asked to do. [This was confirmed in an OSHA Formal Interpretation (August 20, 1991).]*

This volume presents the hazardous materials legislation, regulations, legal issues and background that make up the Hazmat Standard of Care. Also included are the planning process for hazardous materials and the Incident Command System, National Incident Management System, and a look at historical incidents and how they ultimately led to a standard of care.

Elements of the Hazardous Materials Standard of Care

Legal Pit Falls

Law schools across the country are graduating lawyers on a continual basis. Most lawyers in my opinion have good intentions and enter the profession to help people and make a difference in the legal system. There are some, however, who are in the business to make money, lots of money by suing someone as often as they can. Lawyers are at an advantage because they know the law and have a choice of using it to the clients' advantage or to their advantage. My comments come from personal experience in choosing the wrong lawyer out of ignorance. All of that said, public officials including emergency responders are fair game for law suits. Now anyone can sue anyone else for anything at any time. That doesn't, however, mean they will win in court. Nonetheless, when sued most people have to get themselves a lawyer, the cost of which is not cheap. You will spend a great deal of time and money, not to mention the stress to prove your side of the case and hopefully win in court.

There is nothing you can do to prevent a law suit. But you can make sure you will likely win the case by being aware if the standard of care and making sure you do the right things when responding to hazardous materials incidents.

Definition of Liability

"The duty or responsibility to act, provide a service, or perform a duty." Do emergency responders have a responsibility to act, provide a service, or perform a duty? Yes! Whether firefighters, EMS, hazmat or law enforcement, when we began our careers, we accepted those requirements. It doesn't matter if we are career or volunteer; when we became a part of an emergency response organization, we agreed to accept that liability. It doesn't matter whether we knew that at the time or not. Ignorance of the law is no excuse. Emergency responders set out each and every day to do the best job they can for the people they are responsible for serving. In spite of our good intentions, there are those out there waiting for us to screw up. The road to hell is paved with good intentions. Responders must be alert to the fact that regardless of how well intended our actions are, we can get sued. Make sure you are following your department's standard operating procedures (SOPs) and guidelines (SOGs) each and every time we respond to an emergency.

Lawyers are not the only ones we have to look out for. Unfortunately, too often in our society everything is about money. Fire departments have been sued by home owners and insurance companies because they let an insured property be damaged beyond what was expected. Sure there can be mitigating circumstances, but it can cost your department and you a lot of time and money to defend your actions. Let's look at some legal terms that you should be familiar with.

Types of Liability

Vicarious liability: According to the *Legal Dictionary*, "Vicarious liability, sometimes referred to as 'imputed liability', is a legal concept that assigns liability to an individual who did not actually cause the harm, but who has a specific superior legal relationship to the person who did cause the harm. Vicarious liability most commonly comes into play when an employee has acted in a negligent manner for which the employer will be held responsible." Everyone at one time or another makes mistakes. We are human and no one is perfect. As long as negligent acts or omissions were not intentional on your part, you would likely be protected against personnel litigation and responsibility by the protective legal umbrella of your organization.

Individual liability: This means the liability falls on the individual, and they are solely responsible for their own negligent acts or omissions. This umbrella that protected you previously may fold up if your actions are outside of the scope of your job training, standard of care or duties. For example, if you are an EMT you are bound by the standard of care and limited in the services you can perform, regardless of the situation you are in. Hazardous materials personnel are also bound by a standard of care that limits what they can do based upon their level of training and protective equipment available to them. Responders (career or volunteer) take on a big responsibility when they choose the emergency services profession. Intended actions taken that are unlawful or cause harm without legal justification or excuse will likely fold your umbrella, and you would be held personally liable.

Types of Negligence

Negligence: According to the *Legal Dictionary*, negligence means "Failure to exercise the care toward others which a reasonable or prudent person would do in the circumstances, or taking action which such a reasonable person would not. Negligence is accidental." A number of things could cause an emergency responder to be unintentionally negligent. Such as, just forgetting to do something that you would normally do because you were preoccupied with other issues or distracted by something on the incident scene.

Gross negligence: According to the *Legal Dictionary*, gross negligence means "carelessness which is in reckless disregard for the safety or lives of others, and is so great it appears to be a conscious violation of other people's rights to safety. It is more than simple inadvertence, but it is just shy of being intentionally evil." For example, you are an EMT and eating in a restaurant. Another patron of the facility starts choking, and you perform the Heimlich maneuver with no success. You then take a pocket knife and cut into the windpipe so the person can breathe. The person you were trying to help died from loss of blood. You are a medical professional, you acted outside your standard of care for your level of training and you acted intentionally. You could be found grossly negligent. Another example is as follows: as a hazardous materials responder, you arrive on scene and find an injured person lying next to a tanker truck leaking a visible gas vapor. The truck has a poison gas placard and is stenciled "Inhalation Hazard" clearly displayed in your view. You decide to rush in and try to save the downed person. You are overcome and die next to the dead civilian. Your family sues the fire department for wrongful death. In court, the following facts are presented by the defense attorney for the fire department:

- You violated departmental SOPs and SOGs.
- You violated the hazmat standard of care.
- You entered the hot zone without proper training and protective clothing.
- You were found grossly negligent and the suit was thrown out.

Operating outside of a standard of care, failure to operate within departmental policies, and failure to have the proper training and personnel protective equipment have some high costs. Don't find yourself in a situation that may cost you your life and your family's loss of you and benefits. Take hazardous materials regulations and standards seriously. Emergency responders are not exempt from litigation. Operate by gross negligence and you are on your own. Your personnel resources can be attached by a court judgment against you. Do not take chances, the cost is too high, and follow the hazardous materials standard of care (Legal Dictionary).

Four elements of negligence: In order for you to be found negligent for your actions, four factors must be proven. First of all, you must have had a duty to act. If you did not have a duty, you cannot be found to be negligent. Second, you had a duty to act and you failed to perform that duty. Third, there had to be some actual loss to the injured party. Lastly, there has to be a causal connection between the act of the wrong doer and a resulting injury. All of these factors need to be considered before you can be determined to be negligent.

Malfeasance: According to *Webster's Dictionary*, malfeasance means "wrongdoing or misconduct especially by a public official." Misconduct that is willful in terms of hazardous materials response is gross negligence. Failing to perform your duty under the standard of care, according to your job description, follow SOPs or SOGs that could be determined to be malfeasance (Webster's Dictionary).

Sovereign Immunity

According to Web Legal Advise, "Sovereign immunity traces its origins from early English law. Generally, it is the doctrine that the sovereign or government cannot commit a legal wrong and is immune from civil suit or criminal prosecution. For a person individually to be immune to suit, they must be acting as an arm of the government. In many cases, the government has waived this immunity to allow for suits." In order for the governmental unit to be sued, they must give their permission. Oftentimes, states will allow a suit to take place. Employees of a governmental unit in order to be covered by the sovereign immunity "umbrella" they must be free of gross negligence in their actions (Web Legal Advice).

Regulatory Enforcement

Occupational Health and Safety Administration (OSHA) is responsible for the health and safety of workers in the workplace. This includes firefighters, EMS personnel and hazardous materials responders. OSHA Regulation CFR 1910.120(q) applies directly to emergency responders to hazardous materials incidents. Before OSHA was created, there was little if any regulated safety in the workplace, except for union efforts. OSHA regulations were the first regulation of response to hazardous materials incidents and requirements for training of personnel. OSHA also provides inspection of facilities, investigation of fatalities, accidents and workplace complaints.

OSHA Enforcement

Technical violation—minor violation with minimal potential for harm

Serious violation—substantial probability that death or serious physical harm could result, and the employer knew or should have known of the hazard

Willful violation—intentional disregard or plain indifference

Maximum OSHA Violation Penalties 2019

Type of violation	Penalty
Serious	$13,260 per violation
Failure to abate	$13,260 per day beyond the abatement date
Willful or repeated	$132,598 per violation

OSHA is not in the business of making money from violators of its regulations. When compliance cannot be obtained by voluntary compliance, OSHA will levy fines. Citations and fines may also be negotiation through a hearing. Through this negotiation, fines may be eliminated if steps are taken to correct violations. Instead of collecting fines and having them put into the OSHA coffers, they may allow the fined entity to purchase equipment or supplies for local response agencies. Most actions taken by OSHA are civil in nature up to and including court enforcement as a final option. Criminal actions may also be sought in cases where willful or repeated acts have taken place, especially if these acts have resulted in serious injury or loss of life. Criminal actions can result in fines, imprisonment or both (OSHA).

Incidents That Caused Regulatory Change

Unfortunately, our society is often reactive rather than proactive. Monetary issues sometimes outpace safety issues when economic ventures are undertaken. During the 1800s, lack of attention to safety issues was more of a lack of technology and understanding than they were intentional. Nonetheless, the 1800s and early to mid-1900s were filled with disasters of such a magnitude that it forced society to make changes the way it looked at all aspects of safety (Figure 2.10). Regulations and codes were developed, retroactively, but a culture of safety issues in the way we live started to develop. Over the years, the concept evolved into agencies both government and private developing regulations and standards for making our lives and workplaces safer from disaster.

Fires

From the time of the creation of our country, cities across the United States experienced conflagrations that led to formation of organized fire departments and prevention efforts for our communities. Today's modern fire department evolved from bucket brigades over a period of more

Figure 2.10 The 1800s and early to mid-1900s were filled with disasters of such a magnitude that it forced society to make changes the way it looked at all aspects of safety. (Courtesy: Kansas City, MO, Fire Department.)

than 200 years. Sometimes progress in the fire service has been slow and has not kept up with technology. It has been said that the fire service is 200 years of tradition totally unaffected by progress. Along the way fire disasters have given us a push to improve equipment and fire prevention codes.

Great Chicago Fire

On October 10, 1871, during a very long period of hot, dry windy conditions compounded by prevalent wood construction, one of the most famous conflagrations in U.S. history began (Figure 2.11). The origin of the fire was in the area of 137 DeKoven Street, the site of the present-day Chicago Fire Department Training Center. Folklore attributed the ignition of the fire to Mrs. Catherine O'Leary's cow kicked over a lantern. That theory was debunked by the Chicago City Council Committee on Police and Fire nearly 126 years after the fire, and O'Leary was exonerated along with her infamous cow. Flames killed over 300 people, consumed 3.3 square miles of Chicago and left 100,000 people homeless.

The following year, the Chicago City Council mandated the use of fire-resistant materials, such as brick in the construction of future

Figure 2.11 On October 10, 1871, during a very long period of hot, dry windy conditions compounded by prevalent wood construction, one of the most famous conflagrations in U.S. history began, "The Great Chicago Fire."

downtown buildings. The iron industry also developed a way to fire-proof iron columns with porous terra-cotta. This advance was among the factors, including the introduction of steel frame construction, that made possible the invention of the skyscraper in Chicago (Great Chicago Fire).

Baltimore City Fire

On February 7, 1904, another great conflagration consumed portions of downtown in Baltimore, Maryland (Figure 2.12). It destroyed much of central Baltimore and over 1,500 buildings, in an area of two-tenths of a square mile. Unlike Chicago, building construction in Baltimore was largely masonry construction and multiple story buildings. It is considered the third largest conflagration in the United States behind the Great Chicago Fire and the San Francisco Earth Quake Fire. Fire departments responded from as far away as Philadelphia and Washington, DC, with over 1,200 firefighters to battle the blaze. However, it was discovered that there were many different types of hose couplings, and out of town firefighters could not hook up to fire hydrants or couple other departments' hoses together.

The Baltimore Fire led to a national standard hose thread which is still in use today. Additionally, the city adopted a building code, stressing fireproof materials in construction of buildings (Maryland State Archives).

Figure 2.12 On February 7, 1904, another great conflagration consumed portions of downtown in Baltimore, Maryland. It destroyed much of central Baltimore and over 1,500 buildings, in an area of two-tenths of a square mile.

Triangle Shirtwaist Fire, New York City

March 25, 1911, was the deadliest industrial disaster in the history of the city and one of the worst in U.S. history (Figure 2.13). Killed in the fire were 123 women and 23 men. The factory was located on the eighth, ninth and tenth floors of the Asch Building located at 23–29 Washington Place on the corner of Greene and Washington Place, in the Greenwich Village neighborhood of Manhattan. The building still stands today and has been designated a National Historic Landmark and New York City Landmark.

At approximately 4:40 p.m. as the work day was ending, a fire flared up in a scrap bin under one of the cutter's tables at the northeast corner of the eighth floor. The fires alarm was sent by a passerby on Washington Place who saw smoke coming from the eighth floor at 4:45 p.m. Doors to stairwells and exits had been locked by owners, then a common practice to prevent workers from taking unauthorized breaks and to reduce theft. Many trapped by the fire and smoke were forced to jump from high windows to their death. Others were killed by the fire and smoke inhalation.

Figure 2.13 March 25, 1911, was the deadliest industrial disaster in the history of the city and one of the worst in U.S. history. Killed in the fire were 123 women and 23 men.

The Fire Marshal concluded that the likely cause of the fire was the disposal of an unextinguished match or cigarette butt in the scrap bin, full of a month's worth of accumulated cuttings by the time of the fire. The fire led to legislation requiring improved factory safety standards and the formation of the International Ladies' Garment Workers Union (ILGWU), which fought for better working conditions for sweatshop workers (Labor NY.Gov).

Our Lady of the Angels (OLA) Fire, Chicago, IL

On Monday, December 1, 1958, fire broke out at the Our Lady of the Angels School, a K-8 Catholic School, which was located at 909 Avers Avenue in the Humboldt Park neighborhood on Chicago's West side (December 1, 2018, marked the 60th anniversary of the OLA tragedy). The fire started in the basement area of a stairway. There were 1,600 students in the building at the time of the fire. Flames spread quickly through open stairwells and corridors. Ninety-two students and three nuns died when smoke, heat, and toxic gases cut off their escape from corridors and stairways (Figure 2.14). Others were injured when they jumped from second-floor windows which, because the building had a raised basement, were nearly as high as the third floor would be on ground level.

Figure 2.14 Ninety-two students and three nuns died when smoke, heat, and toxic gases cut off their escape from corridors and stairways.

Several delays in the beginning of evacuations and reporting the fire to the fire department occurred. The building's local fire alarm system did not function when first pulled. A teacher came back again a short time later and the system finally sounded the alarm. Since the fire alarm was a local alarm, it was not connected to the fire department. The fire was finally reported to the fire department by telephone when a neighbor saw the smoke coming from the building. Initially, the fire department was sent to the wrong address, which delayed their arrival to the school building.

The cause of the fire was never officially determined. In 1962, a boy who was a student at OLA at the time of the fire, confessed to setting the blaze. He was in the fifth grade and 10 years old at the time of the fire. A family court judge later concluded the evidence was insufficient to substantiate the confession. Officially, the cause of the fire remains unknown. A National Fire Protection Association (NFPA) investigation of the fire blamed the civic authorities (Archdiocese of Chicago) for "housing their children in fire traps".

After the Our Lady of the Angels School fire, Percy Bugbee, the president of the NFPA, said in an interview, "There are no new lessons to be learned from this fire; only old lessons that tragically went unheeded." Sweeping changes in school fire safety regulations were enacted nationwide. Some 16,500 older school buildings in the United States were brought up to code within 1 year of the disaster. Ordinances to strengthen Chicago's fire code and new amendments to the Illinois state fire code were passed. It was estimated that 68% of all U.S. communities inaugurated and completed safety improvements following the Our Lady of the Angels School fire, one of which being an increased number of law-mandated fire drills throughout the academic year. Chicago's City Council passed a law requiring a fire alarm box be installed in front of schools and other public assembly venues. Interior fire alarm systems within buildings had to be connected to the fire alarm box so that the fire department was notified when any alarm was sounded. Another requirement was that all schools where it was deemed vital would have sprinkler systems installed. However, 9 months later, in September 1959, Fire Commissioner Quinn, when interviewed by WNBQ reporter Len O'Connor, admitted that although 400 schools in Chicago at that time had been deemed in critical need of sprinkler systems, only two had actually had sprinklers installed (OLAfire.com, NFPA).

Hazardous Materials

Organized response to hazardous materials is relatively new to emergency response organizations. Jacksonville, Florida, formed the first hazardous materials team in 1978. Martin County, Florida; Houston, Texas; and

Memphis followed over the next year or so. These teams basically pioneered the field of hazardous materials response. There were no regulations or standards, little technology compared to now, they flew by the "seats of their pants". Incidents that had major impacts on fire departments and other emergency response organizations began occurring. The Southwest Boulevard Fire was one of the first hazmat incidents that influenced code changes as a result of the incident outcome.

Southwest Boulevard Fire, Kansas City, Kansas

It was a beautiful but very hot summer day in the Kansas City metropolitan area, sunny with temperatures in the 90s and a south wind of 13 mph. Before the day would end, five Kansas City, MO, firefighters and one civilian would die in an inferno of burning gasoline referred to by KMBC-TV reporter Charles Gray as "when all hell broke loose." Gray, always a strong supporter of the Kansas City Fire Department, called it "one of the darkest days in modern history of Kansas City firefighting." It was the second largest loss of life in Kansas City Fire Department history.

At 8:20 a.m., on August 18, the Kansas City, KS, Fire Department received a report of a fire at the Pyramid Oil Company (Conoco Station) at 2 Southwest Boulevard (Figure 2.15). The fire started on a loading rack

Figure 2.15 At 8:20 a.m., on August 18, the Kansas City, KS, Fire Department received a report of a fire at the Pyramid Oil Company (Conoco Station) at two Southwest Boulevard. (Courtesy of Kansas City, MO, Fire Department.)

at the combination bulk plant and service station in Kansas City, KS, near the Kansas–Missouri state line. Both Kansas City, KS, and Kansas City, MO, fire departments responded to the fire. On the initial alarm, Kansas City, KS, dispatched three pumpers, two ladder trucks, and two district chiefs (a fire apparatus equipped with a pump is referred to as a pumper in the Kansas City area). At 8:35 a.m., two additional pumpers were dispatched from Kansas City, KS.

Additional equipment was called for at 8:45 a.m., including a specially built deluge truck and foam, although foam was not effective on the fire because there was no way to contain the leaking gasoline. The only foam available at the time was protein foam. Although protein foam provides a durable blanket over the surface of a flammable liquid, it spreads slowly and is not as effective as aqueous film-forming foam (AFFF) in fire suppression. In order for the foam to work effectively, the flammable liquid needs to be contained.

At 9:30 a.m., two additional pumpers and off-duty firefighters were summoned from the Kansas City, KS, Fire Department. Chief Edgar Grass of the Kansas City, MO, Fire Department noticed the fire from his office window (the fire was visible for 15 miles in all directions). Knowing the Kansas City, KS, Fire Department was already there and the fire was near the state line, he sent a district chief to investigate. Upon arrival, the district chief immediately requested a first-alarm assignment from the Kansas City, MO, Fire Department. The first alarm was dispatched at 8:33 a.m. and a second alarm was requested at 08:37 a.m., followed by a third at 8:45 a.m., a fourth at 8:54 a.m., a fifth at 8:59 a.m., and a sixth at 10:00 a.m., following the rupture of the tank.

Despite the best efforts of firefighters, the burning gasoline from the leaking fuel extended underneath four 11-by-30-ft cylindrical horizontal storage tanks resting on concrete cradles, each with 21,000 gallons of fuel capacity. Three contained gasoline and one kerosene. From left to right at the fire scene, Tank 1 contained 6,628 gallons of gasoline, Tank 2 contained 15,857 gallons of kerosene, Tank 3 contained 3,000 gallons of gasoline, and Tank 4 contained 15,655 gallons of premium gasoline. All of the tanks failed during the fire, but Tanks 1, 2, and 3 did not leave the concrete cradles they rested in. This lack of movement of the first three tanks may have given firefighters a false sense of security while fighting the fire involving Tank 4.

The tanks began to fail at approximately 10:00 a.m., about 90 min after the fire started. Tank 4 was the last to fail, and when it did, it moved 94 ft from its cradle into Southwest Boulevard through a 13 in. brick wall, spreading burning gasoline and flying bricks in its path. Firefighters with 2½ in. hose lines were just 74 ft from the tank when it ruptured, so their positions were overrun by the tank and burning gasoline that completely crossed Southwest Boulevard (Figure 2.16).

Figure 2.16 Firefighters with 2½ in. hose lines were just 74 ft from the tank when it ruptured, so their positions were overrun by the tank and burning gasoline that completely crossed Southwest Boulevard. (Courtesy of Kansas City, MO, Fire Department.)

As a result of the Southwest Boulevard fire in Kansas City, NFPA's Flammable Liquid Codes were changed to require that all fuel tanks supplying flammable liquids to automobile service stations frequented by the public must be placed underground (*Firehouse Magazine*).

Marshalls Creek, PA, Explosion

Shortly after 4:00 a.m. on Friday, June 26, 1964, the Marshalls Creek Volunteer Fire Company was called to a tractor-trailer fire on the northbound side of Route 209, just southeast of where Regina Farms is now, in Middle Smithfield Township. Just as firefighters arrived to find the burning trailer, which at the time had been abandoned by the driver, the flames spread to the trailer's cargo of nitro-carbo-nitrate (the oxidizing ingredient in fertilizer), partially gelatin dynamite and blasting caps (Figure 2.17). The driver of the truck Albert Koda, 51, of Port Carbon was employed by the American Cyanamid Company. Koda had removed the explosive signs from the trailer and placed them under the seat before unhooking the trailer and going to find an open gas station for help.

The fire caused an explosion that killed firefighters Edward Hines, 42, a welder who had helped start the fire company in 1945; Leonard Mosier, 38, a carpenter, architect, and local rod and gun club member; and Francis Miller, 50, a state highway department employee who also had helped start

Figure 2.17 Just as firefighters arrived to find the burning trailer, which at the time had been abandoned by the driver, the flames spread to the trailer's cargo of nitro-carbo-nitrate (the oxidizing ingredient in fertilizer), partially gelatin dynamite and blasting. (Courtesy of Marshalls Creek Fire Department.)

the fire company and whose brother was chief at the time. Three other people were killed, while two fellow firefighters and eight others were injured. Leaving a large crater visible from the air, the explosion damaged three fire trucks, the Regina Hotel, Middle Smithfield Elementary School, and the Pocono Reptile Farm, sending poisonous snakes flying for miles to the point where it took weeks to find those snakes.

During the 50th anniversary memorial service, Fire Chief Joseph Quaresimo said, "This event we're having marks the anniversary of an incident that was not only tragic, but also changed federal safety regulations to now require all vehicles transporting hazardous or explosive materials to always have warning placards on them. This way, the public is made aware, as well as firefighters responding when one of these vehicles is on fire" (Firehouse.com).

Bhopal, India, December 2–3, 1984, Release of Methyl Isocyanate "A Wake Up Call"

Background

December 2018 marked the 35th anniversary of the massive toxic gas leak from Union Carbide Corporation's (UCC) chemical plant in Bhopal in the state of Madhya Pradesh, India. This incident was a "wake-up call" for the world in terms of illustrating the devastation that chemicals can cause when released into the environment. More than 40 tons of methyl isocyanate (MIC) gas leaked from a pesticide plant owned by the Union Carbide Chemical Company. The company involved in what became the worst industrial accident in history immediately tried to dissociate itself from legal responsibility. Eventually, it reached a settlement with the Indian government through mediation of that country's Supreme Court and accepted moral responsibility. It paid $470 million in compensation, a relatively small amount of based on significant underestimations of the long-term health consequences of exposure and the number of people exposed. The disaster indicated a need for enforceable international standards for environmental safety, preventative strategies to avoid similar accidents, and industrial disaster preparedness.

The Incident

At 11.00 p.m. on December 2, 1984, while most of the 1 million residents of Bhopal slept, an operator at the plant noticed a small leak of MIC gas and increasing pressure inside a storage tank. The vent-gas scrubber, a safety device designed to neutralize toxic discharge from the MIC system, had been turned off 3 weeks prior. Apparently a faulty valve had allowed 1 ton of water for cleaning internal pipes to mix with 40 tons of MIC. A 30 ton refrigeration unit that normally served as a safety component to cool the MIC storage tank had been drained of its coolant for use in another part of the plant. Pressure and heat from the vigorous exothermic reaction in the tank continued to build. The gas flare safety system was out of action and had been for 3 months.

At around 1.00 a.m. on December 3, loud rumbling reverberated around the plant as a safety valve gave way sending a plume of MIC gas into the early morning air. Within hours, the streets of Bhopal were littered with human corpses and the carcasses of buffaloes, cows, dogs, and birds. Local hospitals were soon overwhelmed with the injured, a crisis further compounded by a lack of knowledge of exactly what gas was involved and what its effects were. It became one of the worst chemical disasters in history, and the name Bhopal became synonymous with industrial catastrophe.

Around 12:50 a.m., local time, a Union Carbide employee triggered the plant's alarm system as the concentrations of gas around the plant became difficult to tolerate. Activation of the system triggered two siren alarms: one that sounded inside the plant and the other that directed outward to the public and city of Bhopal. The two siren system had been decoupled from one another in 1982, so that it was possible to leave the factory warning system on while turning off the public system, and this was exactly what was done: the public system sounded briefly at 12:50 a.m. and was quickly turned off, as per company procedure meant to avoid alarming the public around the factory over tiny leaks. Workers, meanwhile, evacuated the Union Carbide plant, traveling upwind.

Bhopal's superintendent of police was informed by telephone, by a town inspector, that residents of the neighborhood of Chola (about 2 km [1.25 miles] from the plant) were fleeing a gas leak at approximately 1:00 a.m. Calls to the Union Carbide plant by police between 1:25 and 2:10 a.m. gave assurances twice that "everything is OK" and, on the last attempt made, "we don't know what has happened, sir." With the lack of timely information exchange between Union Carbide and Bhopal authorities, the city's Hamidia Hospital was first told that the gas leak was suspected to be ammonia, then phosgene. Finally, they received an updated report that it was MIC (rather than "methyl isocyanate"), which hospital staff had never heard of, had no antidote for, and received no immediate information about.

The MIC gas emanating from tank E610 petered out at approximately 2:00 a.m. Fifteen minutes later, the plant's public siren was sounded for an extended period of time, after first having been quickly silenced an hour and a half earlier. Some minutes after the public siren sounded, a Union Carbide employee walked to a police control room to inform them of the leak (their first acknowledgment that one had occurred at all) and that "the leak had been plugged." Most city residents who were exposed to the MIC gas were first made aware of the leak by exposure to the gas itself or by opening their doors to investigate the commotion, rather than having instructed to shelter in place, or to evacuate before the arrival of the gas in the first place.

Over 500,000 people were exposed to the MIC. Initial fatalities were estimated to be 2,259, and the government confirmed 3,787 deaths related to the gas release and 558,125 injuries. Injuries included 38,478 temporary partial injuries and 3,900 severely and permanently disabling injuries. Estimates of the number of people killed in the first few days by the plume from the UCC plant run as high as 10,000, with 15,000–20,000 premature deaths reportedly occurring in the subsequent two decades. Several epidemiological studies conducted soon after the accident showed significant morbidity and increased mortality in the exposed population.

Lessons Learned

The events in Bhopal revealed that expanding industrialization in developing countries without concurrent evolution in safety regulations could have catastrophic consequences. The disaster demonstrated that seemingly local problems of industrial hazards and toxic contamination are often tied to global market dynamics. UCC's Sevin production plant was built in Madhya Pradesh not to avoid environmental regulations in the United States but to exploit the large and growing Indian pesticide market. However, the manner in which the project was executed suggests the existence of a double standard for multinational corporations operating in developing countries.

Enforceable uniform international operating regulations for hazardous industries would have provided a mechanism for significantly improved in safety in Bhopal. Even without enforcement, international standards could provide norms for measuring the performance of individual companies engaged in hazardous activities such as the manufacture of pesticides and other toxic chemicals in India. National governments and international agencies should focus on widely applicable techniques for corporate responsibility and accident prevention as much in the developing world context as in advanced industrial nations. Specifically, prevention should include risk reduction in plant location and design and safety legislation. Local governments clearly cannot allow industrial facilities to be situated within urban areas, regardless of the evolution of land use over time. Industry and government need to bring proper financial support to local communities so they can provide medical and other necessary services to reduce morbidity, mortality, and material loss in the case of industrial accidents. Public health infrastructure was very weak in Bhopal in 1984. Tap water was available for only a few hours a day and was of very poor quality. With no functioning sewage system, untreated human waste was dumped into two nearby lakes, one being a source of drinking water.

The city had four major hospitals, but there was a shortage of physicians and hospital beds. There was also no mass casualty emergency response system in place in the city. Existing public health infrastructure needs to be taken into account when hazardous industries choose sites for manufacturing plants. Future management of industrial development requires that appropriate resources be devoted to advance planning before any disaster occurs. Communities that do not possess infrastructure and technical expertise to respond adequately to such industrial accidents should not be chosen as sites for hazardous industry.

Since 1984

The Bhopal disaster could have changed the nature of the chemical industry and caused a reexamination of the necessity to produce such potentially harmful products in the first place. However, the lessons of acute and chronic effects of exposure to pesticides and their precursors in Bhopal have not changed agricultural practice patterns. An estimated 3 million people per year suffer the consequences of pesticide poisoning with most exposure occurring in the agricultural developing world. It is reported to be the cause of at least 22,000 deaths in India each year.

In the state of Kerala, significant mortality and morbidity have been reported following exposure to Endosulfan, a toxic pesticide whose use continued for 15 years after the events of Bhopal.

Aggressive marketing of asbestos continues in developing countries as a result of restrictions being placed on its use in developed nations due to the well-established link between asbestos products and respiratory diseases. India has become a major consumer, using around 100,000 tons of asbestos per year, 80% of which is imported with Canada being the largest overseas supplier. Mining, production, and use of asbestos in India are very loosely regulated despite the health hazards. Reports have shown morbidity and mortality from asbestos-related disease will continue in India without enforcement of a ban or significantly tighter controls.

Some positive changes were seen following the Bhopal disaster. The British chemical company, Imperial Chemical Industries, whose Indian subsidiary manufactured pesticides, increased attention to health, safety, and environmental issues following the events of December 1984. The subsidiary now spends 30%–40% of their capital expenditures on environmental-related projects. However, they still do not adhere to standards as strict as their parent company in the United Kingdom.

The U.S. chemical giant DuPont learned its lesson of Bhopal in a different way. The company attempted for a decade to export a nylon plant from Richmond, VA, to Goa, India. In its early negotiations with the Indian government, DuPont had sought and won a remarkable clause in its investment agreement that absolved it from all liabilities in case of an accident. But the people of Goa were not willing to acquiesce while an important ecological site was cleared for a heavy polluting industry. After nearly a decade of protesting by Goa's residents, DuPont was forced to scuttle plans there. Chennai was the next proposed site for the plastics plant. The state government there made significantly greater demand on DuPont for concessions on public health and environmental protection. Eventually, these plans were also aborted due to what the company called "financial concerns (International Campaign for Justice)…"

Institute West Virginia Methyl Isocyanate Gas Release

Just 8 months after the Bhopal incident, on August 12, 1985, a small cloud of MIC was released from the Union Carbide in Institute West Virginia. One hundred thirty-five residents were affected by the gas cloud, experiencing eye, throat, and lung irritation. Twenty-eight were hospitalized for treatment. Indeed, people who live in the Kanawha River Valley, which much of the world learned recently is also known as Chemical Valley, have endured a long history of pollution of many kinds. For nearly a century, Chemical Valley was home to the largest concentration of chemical plants in the United States, according to a 2004 history by Nathan Cantrell, published by the West Virginia Historical Society.

Definition of Standard of Care

Legal Standard of care refers to the degree of attentiveness, caution, and prudence that a reasonable person in the circumstances would exercise. Failure to meet the standard is negligence, and the person who fails to meet the standard is liable for any damages caused by such negligence. The standard is not subject to a precise definition and is judged on a case by case basis. Certain standards for professionals are established by practice of similar professionals in their community (Legal Dictionary).

National Fire Academy— the level of competency anticipated or mandated in the provision of a service or duty.

Regulatory Basis for Hazardous Materials Standard of Care

Hazardous Materials Laws and Regulations

Interstate Commerce Commission

The Interstate Commerce Commission (ICC) was a regulatory agency in the United States created by the Interstate Commerce Act of 1887 (Figure 2.18). The agency's original purpose was to regulate railroads (and later trucking), to ensure fair rates, to eliminate rate discrimination and to regulate other aspects of common carriers, including interstate bus lines and telephone companies. Congress expanded ICC authority to regulate other modes of commerce beginning in 1906. The agency was abolished in 1995, and its remaining functions were transferred to the Surface Transportation Board. The commission's five members were appointed by the President with the consent of the U.S. Senate. This was the first independent agency (or so-called Fourth Branch).

Creation

The ICC was established by the Interstate Commerce Act of 1887, which was signed into law by President Grover Cleveland. Creation of the commission was the result of widespread and longstanding anti-railroad agitation. Western farmers, specifically those of the Grange Movement, were the dominant force behind the unrest, but Westerners, especially those in rural areas, generally believed that the railroads possessed

Figure 2.18 The **Interstate Commerce Commission (ICC)** was a regulatory agency in the United States created by the Interstate Commerce Act of 1887.

economic power that they systematically abused. A central issue was race discrimination between similarly situated customers and communities. Other potent issues included alleged attempts by railroads to obtain influence over city and state governments and the widespread practice of granting free transportation in the form of yearly passes to opinion leaders (elected officials, newspaper editors, ministers and so on) so as to dampen any opposition to railroad practices.

Various sections of the Interstate Commerce Act banned "personal discrimination" and required shipping rates to be "just and reasonable." President Cleveland appointed Thomas M. Cooley as the first chairman of the ICC. Cooley had been Dean of the University of Michigan Law School and Chief Justice of the Michigan Supreme Court. Following passage of the 1887 act, the ICC proceeded to set maximum shipping rates for railroads. However, in the late 1890s, several railroads challenged the agency's ratemaking authority in litigation, and the courts severely limited the ICC's powers.

Expansion of ICC Authority

A 1914 cartoon shows railroad companies asking the ICC (depicted as Uncle Sam) for permission to raise rates, while the ghost of a horrified William Henry Vanderbilt looks on (Figure 2.19).

Congress expanded the commission's powers through subsequent legislation. The 1893 Railroad Safety Appliance Act gave the ICC jurisdiction over railroad safety, removing this authority from the states, and this was followed with amendments in 1903 and 1910. The Hepburn Act of 1906 authorized the ICC to set maximum railroad rates and extended the agency's authority to cover bridges, terminals, ferries, sleeping cars, express companies, and oil pipelines. A long-standing controversy was how to interpret language in the Act that banned long haul–short haul

Figure 2.19 A 1914 cartoon shows railroad companies asking the ICC (depicted as Uncle Sam) for permission to raise rates, while the ghost of a horrified William Henry Vanderbilt looks on.

fare discrimination. The Mann–Elkins Act of 1910 addressed this question by strengthening ICC authority over railroad rates. This amendment also expanded the ICC's jurisdiction to include regulation of telephone, telegraph and wireless companies.

The Valuation Act of 1913 required the ICC to organize a Bureau of Valuation that would assess the value of railroad property. This information would be used to set rates.

In 1934, Congress transferred the telecommunications authority to the new Federal Communications Commission. In 1935, Congress passed the

Motor Carrier Act, which extended ICC authority to regulate interstate bus lines and trucking as common carrier.

After much controversy over its history and a slow phasing out, the ICC was terminated by the ICC Termination Act of 1995. Congress passed various deregulation measures in the 1970s and early 1980s which diminished ICC authority, including the Railroad Revitalization and Regulatory Reform Act of 1976, the Motor Carrier Act of 1980 and the Staggers Rail Act of 1980. Senator Fred R. Harris of Oklahoma was a strong supporter of abolishing the commission.

Final Chair Gail McDonald oversaw ICC's remaining functions to a new agency. The Surface Transportation Board (STB), which reviews mergers/acquisitions, rail line abandonments and railroad corporate filings. ICC jurisdiction on rail safety (hours of service rules, equipment and inspection standards) was transferred to the Federal Railroad Administration pursuant to the Federal Railroad Safety Act of 1970.

Motor carriers (bus lines, trucking companies) are now regulated by the Federal Motor Carrier Safety Administration (FMCSA), within the U.S. Department of Transportation (USDOT). Prior to its abolition, the ICC issued identification numbers to motor carriers for which it issued licenses. These identification numbers were generally in the form of "ICC MC-000000." When the ICC was dissolved, the function of licensing interstate motor carriers was transferred to FMCSA. All motor carriers with federal licenses now have a USDOT number such as "USDOT000000."

Federal Water Pollution Control Act Amendments of 1972

Background: Why EPA Was Established

When the U.S. Environmental Protection Agency (EPA) was created some 49 years ago, America had just awakened to the seriousness of its environmental pollution problem. Creation of EPA was part of the response to growing public concern and a grass roots movement to "do something" about the deteriorating conditions of water, air and land. For years, raw sewage, industrial, and feedlot wastes had been discharged into rivers and lakes without regard for the cumulative effect that made our waters unfit for drinking, swimming and boating. Smokestack omissions and automobile exhausts made air pollution so bad in certain communities that some people died and many were hospitalized. The land itself was being polluted by indiscriminate dumping of municipal and industrial wastes and some very toxic chemicals that would later come to the fore when their steel drum containers would rust and leak hazardous materials into soil and aquifers.

For decades, Americans had assumed that air and water were free and plentiful and the industrial community gave little thought to pollution.

Following World War II, however, several developments changed this picture. The United States experienced a vast increase in throwaway packaging—cans, bottles, plastics and paper products—and the introduction into the marketplace of thousands of new synthetic organic chemicals. As a result of this deluge of waste land toxic materials, the earth's automatic, self-cleansing, life support systems became increasingly threatened.

By the 1960s, it was obvious that decisive steps had to be taken to correct this imbalance and to prevent future reoccurrences. People from all walks of life and from every part of the political spectrum were expressing their anxieties. Books such as Rachel Carson's *Silent Spring* became best sellers. Foundations, institutes, clubs, college curricula and corporate departments were formed to understand the problem and to arrive at solutions. Environmental agencies were being created or given added responsibilities in most state governments. It was in this atmosphere that the EPA was created in 1970. EPA was not a carefully crafted well-integrated organization in the beginning. President Nixon, by executive order, "reorganized" the Executive Branch by transferring 15 units from existing organizations into a now independent agency, EPA. Four major government agencies were involved.

It was not an easy birth. Air, solid waste, radiological health, water hygiene, and pesticide tolerance functions and personnel had been transferred from the Department of Health, Education, and Welfare; water quality and pesticide label review came from the Interior Department; radiation protection standards came from the Atomic Energy Commission and the Federal Radiation Council; pesticide registration came from the Department of Agriculture. Employees so transferred were accustomed to four varieties of policies, procedures and administrative practices. It took several years under an able administrator, William D. Ruckelshaus, to bring relative order out of the resulting chaos.

Clean Water Act of 1970 and Amendments
- Established Federal role in environmental emergencies
- Developed Federal regulations for response
- Mechanism for implementation
  a. National Response System (NRS)
  b. National Oil and Hazardous **Substance Contingency Plan (NCP).**
  c. National Response Team (NRT)
  d. Regional Response Teams (RRTs)

Summary of the Clean Water Act 33 U.S.C. §1251 et seq. (1972)
The Clean Water Act (CWA) establishes the basic structure for regulating discharges of pollutants into the waters of the United States and

regulating quality standards for surface waters. The basis of the CWA was enacted in 1948 and was called the Federal Water Pollution Control Act, but the Act was significantly reorganized and expanded in 1972. "Clean Water Act" became the Act's common name with amendments in 1972. Under the CWA, EPA has implemented pollution control programs such as setting wastewater standards for industry. EPA has also developed national water quality criteria recommendations for pollutants in surface waters.

The CWA made it unlawful to discharge any pollutant from a point source into navigable waters, unless a permit was obtained. EPA's National Pollutant Discharge Elimination System (NPDES) permit program controls discharges. Point sources are discrete conveyances such as pipes or man-made ditches. Individual homes that are connected to a municipal system, use a septic system or do not have a surface discharge do not need an NPDES permit; however, industrial, municipal and other facilities must obtain permits if their discharges go directly to surface waters.

Comprehensive Environmental Response, Compensation, and Liability Act (CERCLA) of 1980

- Created the original "Superfund"
- Expanded the role of the **National Response Team**
- Placed additional emphasis on emergency response
- Established the principle of "the spiller pays"

Superfund: CERCLA Overview

The Comprehensive Environmental Response, Compensation, and Liability Act (CERCLA), commonly known as Superfund, was enacted by Congress on December 11, 1980. This law created a tax on the chemical and petroleum industries and provided broad Federal authority to respond directly to releases or threatened releases of hazardous substances that may endanger public health or the environment. Over 5 years, $1.6 billion was collected and the tax went to a trust fund for cleaning up abandoned or uncontrolled hazardous waste sites. The Comprehensive Environmental Response, Compensation, and Liability Act of 1980 (CERCLA):

- Established prohibitions and requirements concerning closed and abandoned hazardous waste sites
- Provided for liability of persons responsible for releases of hazardous waste at these sites
- Established a trust fund to provide for cleanup when no responsible party could be identified

The law authorizes two kinds of response actions:

- Short-term removals, where actions may be taken to address releases or threatened releases requiring prompt response
- Long-term remedial response actions, which permanently and significantly reduce the dangers associated with releases or threats of releases of hazardous substances that are serious, but not immediately life threatening. These actions can be conducted only at sites listed on EPA's National Priorities List

CERCLA also enabled the revision of the National Contingency Plan (NCP). The NCP provided the guidelines and procedures needed to respond to releases and threatened releases of hazardous substances, pollutants or contaminants. The NCP also established the National Priorities List. CERCLA was amended by the Superfund Amendments and Reauthorization Act (SARA) on October 17, 1986.

The SARA aka Emergency Planning and Community Right-to-Know Act (EPCRA)

SARA
- Four Titles
  a. Title I—Response and Liability Provisions
  b. Title II—Miscellaneous Hazardous Waste Provisions
  c. Title III—Emergency Planning and Community Right-to-Know (EPCRA)
  d. Title IV—Radon Gas and Indoor Air Quality Research

Title I—Section 126
a. Mandates OSHA and EPA regulations on Hazardous Waste Operations and Emergency Response (HAZWOPER)
b. Is the only section in Title I that applies to emergency response

Title III—EPCRA
a. Planning and reporting mandate
b. Addresses airborne releases of extremely hazardous substances (EHSs)—established by EPA) only

The **Superfund Amendments and Reauthorization Act amended the Comprehensive Environmental Response, Compensation, and Liability Act of 1980 (CERCLA) on October 17, 1986. The Superfund Amendments and Reauthorization Act of 1986 (**SARA reflected EPA's experience in administering the complex Superfund program during its first 6 years and made several important changes and additions to the program.

SARA

- stressed the importance of permanent remedies and innovative treatment technologies in cleaning up hazardous waste sites.
- required Superfund actions to consider the standards and requirements found in other state and Federal environmental laws and regulations.
- provided new enforcement authorities and settlement tools.
- increased State involvement in every phase of the Superfund program.
- increased the focus on human health problems posed by hazardous waste sites.
- encouraged greater citizen participation in making decisions on how sites should be cleaned up.
- increased the size of the trust fund to $8.5 billion.

SARA also required EPA to revise the Hazard Ranking System to ensure that it accurately assessed the relative degree of risk to human health and the environment posed by uncontrolled hazardous waste sites that may be placed on the National Priorities List(NPL).

EPCRA Overview

The **Emergency Planning and Community Right-to-Know Act (EPCRA)** was passed in 1986 in response to concerns regarding the environmental and safety hazards posed by the storage and handling of toxic chemicals. These concerns were triggered by the 1984 disaster in Bhopal, India, caused by an accidental release of methyl isocyanate. The release killed or severely injured more than 2,000 people.

EPCRA Fact Sheet

What are SERCs, TERCS and LEPCs? The governor of each state designated a State Emergency Response Commission (SERC). The SERCs, in turn, designated about 3,500 local emergency planning districts and appointed Local Emergency Planning Committees (LEPCs) for each district (Figure 2.20). The SERC supervises and coordinates the activities of the LEPCs, establishes procedures for receiving and processing public requests for information collected under EPCRA and reviews local emergency response plans. The Chief Executive Office of the Tribe appoints the Tribal Emergency Response Commissions (TERCs).

TERCs have the same responsibilities as the SERCs. The LEPC membership must include, at a minimum, local officials including police, fire, civil defense, public health, transportation and environmental professionals, as well as representatives of facilities subject to the emergency planning requirements, community groups and the media. The LEPCs

Figure 2.20 The SERCs, in turn, designated about 3,500 local emergency planning districts and appointed **Local Emergency Planning Committees (LEPCs)** for each district.

must develop an emergency response plan, review it at least annually and provide information about chemicals in the community to citizens.

Regulations implementing EPCRA are codified in Title 40 of the Code of Federal Regulations (CFR), parts 350–372. The chemicals covered by each of the sections are different, as are the quantities that trigger reporting. Emergency response plans contain information that community officials can use at the time of a chemical accident. Community emergency response plans for chemical accidents were developed under Section 303.

Planning activities of LEPCs and facilities initially focused on, but were not limited to, the 406 extremely hazardous substances EHSs listed by EPA in 1987. These substances are currently a part of EPA's List of Lists (March 2015) which contain the following chemical lists regulated by EPA:

- EPCRA Section 302 Extremely Hazardous Substances
- CERCLA Hazardous Substances
- EPCRA Section 313 Toxic Chemicals
- CAA 112(r) Regulated Chemicals for Accidental Release Prevention

The list includes the threshold planning quantities (minimum limits) for each substance. Any facility that has EHS at or above its threshold planning quantity must notify the State Emergency Response Commission (SERC) or the Tribal Emergency Response Commission (TERC) and Local Emergency Planning Committee (LEPC) within 60 days after they first receive a shipment or produce the substance on -site.

What Are the Emergency Notification Requirements (Section 304)?
Facilities must immediately notify the LEPC and the SERC or the TERC if there is a release into the environment of a hazardous substance that is equal to or exceeds the minimum reportable quantity set in the regulations (Figure 2.21). They are also required to notify the National Response Center (NRC). This requirement covers the 355 extremely hazardous substances, as well as the more than 700 hazardous substances subject to the emergency notification requirements under CERCLA Section 103(a) (40 CFR 302.4). Some chemicals are common to both lists. Initial notification can be made by telephone, radio, or in person. Emergency notification requirements involving transportation incidents can be met by dialing 911 or, in the absence of a 911 emergency number, calling the operator. This emergency notification needs to include the following:

- The chemical name
- An indication of whether it is an extremely hazardous substance
- An estimate of the quantity released into the environment
- The time and duration of the release
- Whether the release occurred into air, water and/or land

Figure 2.21 Facilities must immediately notify the LEPC and the SERC or the TERC if there is a release into the environment of a hazardous substance that is equal to or exceeds the minimum reportable quantity set in the regulations. (Courtesy of Houston Fire Department.)

- Any known or anticipated acute or chronic health risks associated with the emergency and, where necessary, advice regarding medical attention for exposed individuals
- Proper precautions, such as evacuation or sheltering in place
- Name and telephone number of contact person

A written follow-up notice must be submitted to the SERC or the TERC and LEPC as soon as practicable after the release. The follow-up notice must update information included in the initial notice and provide information on actual response actions taken and advice regarding medical attention necessary for citizens exposed.

What Are the Community Right-to-Know Requirements (Sections 311 and 312)?

Under Occupational Safety and Health Administration (OSHA) regulations, employers must maintain a material safety data sheet (MSDS) for any hazardous chemicals stored or used in the workplace.

Approximately 500,000 products are required to have MSDSs. Section 311 requires facilities that have MSDSs for chemicals held above certain threshold quantities to submit either copies of their MSDSs or a list of these chemicals to the SERC or TERC, LEPC and local fire department. If the facility owner or operator chooses to submit a list of chemicals, the list must include the chemical or common name of each substance and must identify the applicable hazard categories. These hazard categories are as follows:

Physical Hazards
- Flammable (gases, aerosols, liquids or solids)
- Gas under pressure
- Explosive self-heating pyrophoric (liquid or solid)
- Oxidizer (liquid, solid or gas)
- Organic peroxide self-reactive in contact with water emits flammable gas
- Corrosive to metal Hazard Not Otherwise Classified (HNOC)

If a list is submitted, the facility must submit a copy of the MSDSs for any chemical on the list upon request by the LEPC. Facilities that start using a hazardous chemical or increase the quantity to exceed the thresholds must submit MSDSs or a list of MSDSs for chemicals within 3 months after they become covered. Facilities must provide a revised MSDS to update the original MSDS or list if significant new information is discovered about the hazardous chemical. Facilities covered by Section 311 must submit annually an Emergency and Hazardous Chemical Inventory Form to the LEPC, SERC or TERC, and the local fire department as required under Section 312. Facilities provide either a Tier I or a Tier II inventory form. Tier I inventory form includes the following aggregate information for each applicable hazard category:

- An estimate (in ranges) of the maximum amount of hazardous chemicals for each category present at the facility at any time during the preceding calendar year
- An estimate (in ranges) of the average daily amount of hazardous chemicals in each category and,

Health Hazards
- Carcinogenicity acute toxicity (any route of exposure)
- Reproductive toxicity
- Skin corrosion or irritation respiratory **or**
- Skin sensitization serious eye damage or
- Eye irritation
- Specific target organ toxicity (single or repeated exposure)
- Germ cell mutagenicity
- Aspiration **Hazard Not Otherwise Classified (HNOC)**

The general location of hazardous chemicals in each category. The Tier II inventory form contains basically the same information as the Tier I, but it must list the specific chemicals. Tier II inventory form provides the following information for each chemical:

- The chemical name or the common name as indicated on the MSDS
- An estimate (in ranges) of the maximum amount of the chemical present at any time during the preceding calendar year and the average daily amount
- A brief description of the manner of storage of the chemical
- The location of the chemical at the facility
- An indication of whether the owner elects to withhold location information from disclosure to the public

Many states now require Tier II inventory form or the state equivalent including electronic reporting under state law. Section 312 information must be submitted on or before March 1 each year for information on chemicals present at the facility in the previous year. The information submitted under Sections 311 and 312 is available to the public from LEPCs and SERCs or TERCs.

What Is the Toxics Release Inventory (Section 313)?
Section 313 of EPCRA established the Toxics Release Inventory (TRI).

TRI tracks the management of certain toxic chemicals that pose a threat to human health and the environment. Facilities in different industry sectors must annually report how much of each chemical they managed through recycling, energy recovery, treatment and environmental releases. TRI reporting forms must be submitted to EPA and the appropriate state or tribe by July 1 of each year. These forms cover environmental releases and other management of toxic chemicals that occurred during the previous calendar year. The information submitted by facilities is compiled in the TRI and made available to the public through the TRI website: www.epa.gov/toxics-release inventory-tri-program. TRI helps support informed decision-making by industry, government, nongovernmental organizations and the public. TRI includes information about the following:

- On-site releases (including disposal) of toxic chemicals to air, surface water and land
- On-site recycling, treatment and energy recovery associated with TRI chemicals
- Off-site transfers of toxic chemicals from TRI facilities to other locations
- Pollution prevention activities at facilities
- Releases of lead, mercury, dioxin and other persistent, bioaccumulative and toxic (PBT) chemicals
- Facilities in a variety of industry sectors (including manufacturing, metal mining and electric power generation) and some federal facilities. A complete list of covered facility sectors is available online: www.epa.gov/toxics-releaseinventory-tri-program/my-facilitys-six-digit-code-tri-coverd-industry. Some of the ways TRI data can be used include the following:
  - Identifying sources of toxic chemical releases
  - Beginning to analyze potential toxic chemical hazards to human health and the environment and
  - Encouraging pollution prevention at facilities

Table 1: EPCRA Chemicals and Reporting Thresholds
Chemicals Covered
- Section 302: 355 Extremely Hazardous Substances
- Section 304: 1,000 substances
- Sections 311/312: approximately 800,000 hazardous chemicals
- Section 313: 650 toxic chemicals and categories

Thresholds
- Threshold planning quantity, 110,000 lb on-site at any one time
- Reportable quantity, 15,000 lb, released in a 24 h period
- 500 pounds or threshold planning quantity (TPQ), whichever is less for EHSs

- Gasoline greater than or equal to 75,000 gallons (all grades combined)*
- Diesel greater than or equal to 100,000 gallons (all grades combined)*
- 10,000 pounds for all other hazardous chemicals
- 25,000 pounds per year manufactured or processed; 10,000 pounds a year otherwise used; persistent bio accumulative toxics have lower thresholds

*These thresholds are only applicable for gasoline and diesel present at retail gas stations in tank(s) entirely underground and was in compliance at all times during the preceding calendar year with all applicable Underground Storage Tank (UST) requirements at 40 CFR part 280 or requirements of the state UST program approved by the Agency under 40 CFR part 281.

What Else Does EPCRA Require?

Trade secrets. EPCRA Section 322 allows facilities to file trade secrets in their reports under EPCRA Sections 303, 311, 312 and 313. Only the specific chemical identity may be claimed as a trade secret, though a generic class for the chemical must be provided. The criteria a facility must meet to claim a chemical identity as a trade secret are in 40 CFR part 350. A facility cannot claim trade secrets under EPCRA Section 304. Even if specific chemical identity information can be legally withheld from the public, EPCRA Section 323 allows the information to be disclosed to health professionals who need the information for diagnostic and treatment purposes or local health officials who need the information for prevention and treatment. In nonemergency cases, the health professional must sign a confidentiality agreement with the facility and provide a written statement of need. During a medical emergency, the health professional may obtain the specific chemical identity from the facility for treatment. Any person may challenge trade secret claims by petitioning EPA. The Agency must then review the claim and rule on its validity.

EPCRA penalties. EPCRA Section 325 allows civil and administrative penalties ranging from $21,916 to $1,643,671 per violation per day when facilities fail to comply with the reporting requirements. Criminal penalties up to $50,000 or 5 years in prison apply to any person who knowingly and willfully fails to provide emergency release notification. Penalties of not more than $20,000 and/or up to 1 year in prison apply to any person who knowingly and willfully discloses any information entitled to protection as a trade secret.

Citizens suits. EPCRA Section 326 allows citizens to initiate civil actions against EPA, SERCs and the owner or operator of a facility for failure to meet the EPCRA requirements. A SERC or TERC, LEPC and state or local government may institute actions against facility owner or operator for failure to comply with EPCRA requirements. In addition, states may sue EPA for failure to provide trade secret information.

To reduce the likelihood of such a disaster in the United States, Congress imposed requirements for federal, state and local governments, tribes and industry. These requirements covered emergency planning and "Community Right-to-Know" reporting on hazardous and toxic chemicals. The Community Right-to-Know provisions help increase the public's knowledge and access to information on chemicals at individual facilities, their uses and releases into the environment. States and communities, working with facilities, can use the information to improve chemical safety and protect public health and the environment.

Key Provisions of the Emergency Planning and Community Right-to-Know Act

Sections 301–303. Emergency Planning—Local governments are required to prepare chemical emergency response plans and to review plans at least annually. State governments are required to oversee and coordinate local planning efforts. Facilities that maintain Extremely Hazardous Substances EHSs on-site in quantities greater than their corresponding threshold planning quantities TPQs must cooperate in emergency plan preparation.

Section 304. Emergency Notification—Facilities must immediately report accidental releases of EHSs and "hazardous substances" defined under the Comprehensive Environmental Response, Compensation, and Liability Act CERCLA. Any releases of these substances in quantities greater than their corresponding reportable quantities must be reported to state and local officials. See also Continuous Release Reporting.

Sections 311 and 312. Community Right-to-Know Requirements— Facilities handling or storing any hazardous chemicals must submit Material Safety Data Sheets MSDSs (or safety data sheets [SDSs]) to state and local officials and local fire departments (Figure 2.22). Hazardous chemicals are defined under the Occupational Safety and Health Act and its implementing regulations. MSDSs (or SDSs) describe the properties and health effects of these chemicals. Facilities must also submit an inventory form for these chemicals to state and local officials and local fire departments.

Section 313. Toxics Release Inventory TRI—Facilities must complete and submit a toxic chemical release inventory form (Form R) annually. Form R must be submitted for each of the over 600 TRI chemicals that are manufactured or otherwise used above the applicable threshold quantities. CERCLA was amended by the Superfund Amendments and Reauthorization Act (SARA) on October 17, 1986.

Figure 2.22 Facilities handling or storing any hazardous chemicals must submit **Material Safety Data Sheets (MSDSs)** (or Safety Data Sheets, SDSs) to state and local officials and local fire departments.

Hazardous Materials Transportation Uniform Safety Act of 1990 (HMTUSA)

- Now called the Hazardous Materials Transportation Act (HMTA)
- Integrates U.S. with international requirements
- Hazard class/division system, packaging
- Placarding labeling
- Commercial driver's hazmat licensing
- Planning and training funding mechanism
- Hazardous Materials Emergency Preparedness (HMEP) national curriculum guidelines

In 1990, Congress enacted the Hazardous Materials Transportation Uniform Safety Act (HMTUSA) to clarify the maze of conflicting state, local and federal regulations. Like the HMTA, the HMTUSA requires the Secretary of Transportation to promulgate regulations for the safe transport of hazardous material in intrastate, interstate and foreign commerce (Figure 2.23). The Secretary also retains authority to designate materials as hazardous when they pose unreasonable risks to health, safety or property.

The statute includes provisions to encourage uniformity among different state and local highway routing regulations, to develop criteria for

Figure 2.23 Like the HMTA, the HMTUSA requires the Secretary of Transportation to promulgate regulations for the safe transport of hazardous material in intrastate, interstate and foreign commerce.

the issuance of federal permits to motor carriers of hazardous materials and to regulate the transport of radioactive materials.

National Oil and Hazardous Substances
Pollution Contingency Plan (NCP)

The National Oil and Hazardous Substances Pollution Contingency Plan, more commonly called the National Contingency Plan or NCP, is the federal government's blueprint for responding to both oil spills and hazardous substance releases (Figure 2.24). The **National Contingency Plan,** is the result of our country's efforts to develop a national response capability and promote overall coordination among the hierarchy of responders and contingency plans. The first **National Contingency Plan** was developed and published in 1968 in response to a massive oil spill from the oil tanker SS Torrey Canyon off the coast of England the year before (Figure 2.25).

More than 37 million gallons of crude oil spilled into the water, causing massive environmental damage. To avoid the problems faced by response officials involved in this incident, U.S. officials developed a coordinated approach to cope with potential spills in U.S. waters. The 1968 Plan provided the first comprehensive system of accident reporting, spill containment and cleanup, and established a response headquarters, a national reaction team and regional reaction teams (precursors to the current **National Response Team** and **Regional Response Teams**).

Figure 2.24 The National Oil and Hazardous Substances Pollution Contingency Plan, more commonly called the National Contingency Plan or NCP, is the federal government's blueprint for responding to both oil spills and hazardous substance releases.

Figure 2.25 The first **National Contingency Plan** was developed and published in 1968 in response to a massive oil spill from the oil tanker SS Torrey Canyon off the coast of England the year before.

Congress has broadened the scope of the **National Contingency Plan** over the years. As required by the **Clean Water Act** of 1972, the NCP was revised the following year to include a framework for responding to hazardous substance spills as well as oil discharges. Following the passage of Superfund legislation in 1980, the NCP was broadened to cover releases at hazardous waste sites requiring emergency removal actions. Over the years, additional revisions have been made to the NCP to keep pace with the enactment of legislation. The latest revisions to the NCP were finalized in 1994 to reflect the oil spill provisions of the Oil Pollution Act of 1990.

National Response Center
United States Coast Guard (USCG) coordinates response and helps run the **National Response Center**, which became operational in August of 1974 at **US Coast Guard** Headquarters in Washington, DC, for the reporting and coordination of response to pollution by oil and hazardous substances. The **National Response Center** is a part of the federally established **National Response System** and staffed 24 h a day by the **U.S. Coast Guard**. It is the designated federal point of contact for reporting all oil, chemical, radiological, biological and etiological discharges into the environment, anywhere in the United States and its territories.

The NRC also takes maritime reports of suspicious activity and security breaches within the waters of the United States and its territories. Reports to the NRC activate the **National Contingency Plan** and the federal government's response capabilities. It is the responsibility of the NRC staff to notify the predesignated On-Scene Coordinator assigned to the area of the incident and to collect available information on the size and nature of the release, the facility or vessel involved and the party(ies) responsible for the release. The NRC maintains reports of all releases and spills in a national database. You may contact (800) 424-8802 to report oil and hazardous chemical spills.

NRC also has communicators on duty much like Chemical Transportation Emergency Center (CHEMTREC). Like CHEMTREC is the entry point for resources in the private sector chemical industry, NRC is the entry point into Federal Government resources during a hazardous materials emergency. They can provide chemical information using the computerized version of OM-TADS, the Coast Guard's chemical database. If you want to make use of the EPA $25,000 reimbursement program, you must report your incident to the NRC. Reimbursement will not start until that call is made. So it is necessary to have someone report your incident immediately after arrival on scene if it looks like you will be creating expenses on your department. Spillers are responsible for paying for costs of hazmat response and clean-up. You would bill them first. **If they** cannot or do not pay, then if you reported your incident to the NRC you can participate in the EPA program.

> **Author's Note:** *While taking a class and later teaching a class at the National Fire Academy, we took field trips to CHEMTREC and to the National Response Center. We could not take photos inside the NRC for security reasons but learned a lot about how they operate. When we entered the center, I noticed they had televisions with major networks. I told them I thought it was nice that they could watch television while on duty. I was told they learned more about incidents occurring by watching television than through phone calls. One of the communicators told us that sometimes people just called them to talk. One of the reasons why we could not take photos was that screens were displayed showing the locations of Coast Guard ships and other assets.*

National Response Team (NRT)

Response planning and coordination is accomplished at the federal level through the **U.S. National Response Team** an interagency group co-chaired by the EPA and the USCG (Figure 2.26). Although the NRT does

Figure 2.26 Response planning and coordination is accomplished at the federal level through the **U.S. National Response Team (NRT),** an interagency group co-chaired by the EPA and the **U.S. Coast Guard.**

not respond directly to incidents, it is responsible for **three** major activities related to managing responses:

- The NRT is responsible for distributing technical, financial and operational information about hazardous substance releases and oil spills to all members of the team. Standing committees of the NRT and the topics that are addressed include the following:
- **Response Committee**, chaired by EPA, addresses issues such as response operations, technology employment during response, operational safety and interagency facilitation of response issues (e.g., customs on transboundary issues). Response-specific national policy/program coordination and capacity building also reside in this committee.
- **Preparedness Committee**, chaired by the **U.S. Coast Guard**, addresses issues such as preparedness training, monitoring exercises/drills, planning guidance, planning interoperability and planning consistency issues. Preparedness-specific national policy/ program coordination and capacity building also reside in this committee.
- **Science and Technology Committee**, chaired by EPA and the National Oceanic and Atmospheric Administration in alternating years, provides national coordination on issues that parallel those addressed by the Scientific Support Coordinator on an incident-by-incident basis. The focus of this committee is to identify technology and mechanisms to apply and enhance operational response. The committee monitors research and development of response technologies and provides relevant information to the RRTs and other members of the **National Response System** to assist in the use of such technologies.

Planning for Emergencies   The NRT ensures that the roles of federal agencies during an emergency response are clearly outlined in the **National Contingency Plan** (see the National Oil and Hazardous Substances Pollution Contingency Plan Overview). After a major incident, the effectiveness of the response is carefully assessed by the NRT. The NRT may use information gathered from the assessment to make recommendations for improving the **National Contingency Plan** and the **National Response System**. The NRT may be asked to help RRTs develop regional contingency plans. It also reviews these plans to determine whether they comply with federal policies on emergency response.

Training for Emergencies   Training is the key to the federal strategy to prepare for oil spills or hazardous substance releases. Although most training is performed by state and local personnel, the NRT develops training courses and programs, coordinates federal training efforts and provides information to regional, state and local officials about training needs and courses.

Supporting Regional Response Teams

The NRT supports RRTs by reviewing regional or area contingency plans to maintain consistency with national policies on emergency response. **The NRT** also supports RRTs by monitoring and assessing RRT effectiveness during an incident. The NRT can ask an RRT to focus on specific lessons learned from a particular incident and to share those lessons with other members of the **National Response System.** This allows the RRTs to improve their own regional contingency plans while helping solve problems that might occur elsewhere within the **National Response System**.

Regional Response Teams

There are 13 **Regional Response Teams (RRTs) in** the United States, each representing a particular geographic region (including the Caribbean and the Pacific Basin) (Figure 2.26A). RRTs are composed of representatives from field offices of the federal agencies that make up the **National Response Team (NRT),** as well as state representatives. The four major responsibilities of RRTs are response, planning, training and coordination.

Figure 2.26A There are 13 **Regional Response Teams (RRTs)** in the United States, each representing a particular geographic region (including the Caribbean and the Pacific Basin).

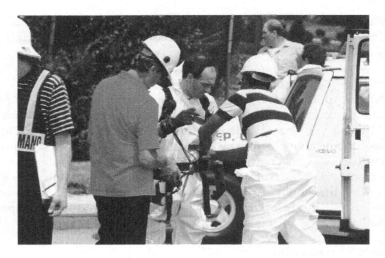

Figure 2.26B As with the NRT, RRT members do not respond directly to releases or spills, but may be called upon to provide technical advice, equipment or manpower to assist with a response.

Response RRTs provide a forum for federal agency field offices and state agencies to exchange information about their abilities to respond to **On-Scene Coordinator's (OSC's)** requests for assistance. As with the NRT, RRT members do not respond directly to releases or spills, but may be called upon to provide technical advice, equipment or manpower to assist with a response (Figure 2.26B).

Planning Each RRT develops a regional contingency plan to ensure that the roles of federal and state agencies during an actual incident are clear. Following an incident, the RRT reviews the OSC's reports to identify problems with the region's response to the incident and improves the plan as necessary.

Training Federal agencies that are members of the RRTs provide simulation exercises of regional plans. These exercises test the ability of federal, state and local agencies to coordinate their emergency response activities. Any major problems identified as a result of these exercises may be addressed and changed in the regional contingency plan so the same problems do not arise during an actual incident.

Coordination The RRTs identify available resources from each federal agency and state within their regions. Such resources include equipment, guidance, training and technical expertise for dealing with chemical releases or oil spills. When there are too few resources in a region, the

RRT can request assistance from federal or state authorities to ensure that sufficient resources will be available during an incident. This coordination by the RRTs assures that resources are used as wisely as possible, and that no region is lacking what it needs to protect human health and the environment.

Responding to an Incident

First Line of Defense When a release or spill occurs, the first line of defense is provided by the following:

- The company responsible for the release
- Its response contractors
- Local fire and police departments
- Local emergency response personnel

If needed, a variety of state agencies stand ready to support, assist or take over response operations if an incident is beyond local capabilities. In some cases, local governments or Indian tribes conduct temporary emergency measures, but do not have emergency response funds budgeted to cover response costs.

Federal Involvement

If the amount of a hazardous substance release or oil spill exceeds the established reporting trigger, the organization responsible for the release or spill is required by law to notify the federal government's **National Response Center (NRC)**. Once a report is made, the NRC immediately notifies a predesignated EPA or **U.S. Coast Guard On-Scene Coordinator (OSC),** based on the location of the spill. The OSC determines the status of the local response and monitors the situation to determine whether, or how much, federal involvement is necessary. It is the OSC's job to ensure that the clean-up, whether accomplished by industry, local, state or federal officials, is appropriate and timely, and minimizes human and environmental damage.

The federal OSC will take command of the response in the following situations:

- If the party responsible for the chemical release or oil spill is unknown or not cooperative
- If the OSC determines that the spill or release is beyond the capacity of the company, local or state responders to manage
- For oil spills, if the incident is determined to present a substantial threat to public health or welfare due to the size or character of the spill

The OSC may request additional support to respond to a release or spill, such as additional contractors, technical support from EPA's special teams, or scientific support coordinators from EPA or the National Oceanic and Atmospheric Administration. The OSC also may seek support from the **Regional Response Team** to access special expertise or to provide additional logistical support. In addition, the **National Response Team** stands ready to provide backup policy and logistical support to the OSC and the RRT during an incident. The **National Response System (NRS)** flowchart provides a quick reference for how additional resources are brought into the response.

The federal government will remain involved at the site following response actions to undertake a number of activities, including assessing damages, supporting restoration efforts, recovering response costs from the parties responsible for the spill and, if necessary, enforcing the liability and penalty provisions of the **Clean Water Act**, as amended by the Oil Pollution Act of 1990.

EPA's Role in Emergency Response—Special Teams

Environmental Response Team

The Environmental Response Team (ERT) is a group of EPA technical experts who provide around-the-clock assistance at the scene of hazardous substance releases (Figure 2.27). ERT offers expertise in

Figure 2.27 The **Environmental Response Team (ERT)** is a group of EPA technical experts who provide around-the-clock assistance at the scene of hazardous substance releases.

such areas as treatment, biology, chemistry, hydrology, geology and engineering. It can provide support to the full range of emergency response actions, including unusual or complex emergency incidents. In such cases, ERT can bring in special equipment and experienced responders, and can provide the OSC or lead responder with experience and advice.

Radiological Emergency Response Team

The Radiological Emergency Response Team (RERT) is a specialized unit that responds to emergencies requiring the clean-up of radioactive substances. RERT provides on-site and lab-based radiation risk monitoring services.

Chemical, Biological, Radiological, and Nuclear Consequence Management Advisory Division

The Chemical, Biological, Radiological, and Nuclear Consequence Management Advisory Division (CBRN CMAD) provides scientific support and technical expertise for decontamination of the following:

- Buildings
- Building contents
- Public infrastructure
- Agriculture
- Associated environmental media

CMAD provides specialized expertise, such as biochemistry, microbiology and medicine, health physics, toxicology, HVAC engineering and industrial hygiene. **CMAD** is available to assist local, national and international agencies supporting hazardous substance response and remedial operations, including nationally significant incidents.

National Criminal Enforcement Response Team

The Office of Criminal Enforcement, Forensics, and Training's National Criminal Enforcement Response Team (NCERT) supports environmental crime investigations involving chemical, biological or radiological releases to the environment. NCERT's specially trained law enforcement officers

- Collect forensic evidence within contaminated zones.
- Serve as law enforcement liaisons with other law enforcement agencies; **and**
- Provide protective escorts to EPA's OSCs, contractors and other EPA special teams during national emergencies.

Summary of the Clean Air Act

The Clean Air Act (CAA) is the comprehensive federal law that regulates air emissions from stationary and mobile sources. Among other things, this law authorizes EPA to establish National Ambient Air Quality Standards (NAAQS) to protect public health and public welfare and to regulate emissions of hazardous air pollutants. One of the goals of the Act was to set and achieve NAAQS in every state by 1975 in order to address the public health and welfare risks posed by certain widespread air pollutants. The setting of these pollutant standards was coupled with directing the states to develop state implementation plans (SIPs), applicable to appropriate industrial sources in the state, in order to achieve these standards. The Act was amended in 1977 and 1990 primarily to set new goals (dates) for achieving attainment of NAAQS since many areas of the country had failed to meet the deadlines.

Clean Air Act

- Amends the previous **Clean Air Act**.
- Mandates new Federal focus on chemical accident prevention.
- Designed to complement and support
  a. SARA Title III (EPCRA) planning and information requirements.
  b. OSHA 29 CFR 1910.119. Process Safety Management.
- Involves the development of Risk Management Plans **(RMPS).**

The legal authority for federal programs regarding air pollution control is based on the 1990 Clean Air Act Amendments (1990 CAAA). These are the latest in a series of amendments made to the **Clean Air Act (CAA)**. This legislation modified and extended federal legal authority provided by the earlier **Clean Air Acts** of 1963 and 1970.

The Air Pollution Control Act of 1955 was the first federal legislation involving air pollution. This Act provided funds for federal research in air pollution. The **Clean Air Act** of 1963 was the first federal legislation regarding air pollution *control*. It established a federal program within the U.S. Public Health Service and authorized research into techniques for monitoring and controlling air pollution. In 1967, the Air Quality Act was enacted in order to expand federal government activities. In accordance with this law, enforcement proceedings were initiated in areas subject to interstate air pollution transport. As part of these proceedings, the federal government for the first time conducted extensive ambient monitoring studies and stationary source inspections. The Air Quality Act of 1967 also authorized expanded studies of air pollutant emission inventories, ambient monitoring techniques and control techniques.

Clean Air Act of 1970

The enactment of the **Clean Air Act** of 1970 (1970 CAA) resulted in a major shift in the federal government's role in air pollution control. This legislation authorized the development of comprehensive federal and state regulations to limit emissions from both stationary (industrial) sources and mobile sources. Four major regulatory programs affecting stationary sources were initiated: the National Ambient Air Quality Standards NAAQS (pronounced "knacks"), State Implementation Plans SIPs, New Source Performance Standards (NSPS) and National Emission Standards for Hazardous Air Pollutants (NESHAPs).

Furthermore, the enforcement authority was substantially expanded. The adoption of this very important legislation occurred at approximately the same time as the National Environmental Policy Act that established the **U.S. Environmental Protection Agency (EPA)**. The EPA was created on December 2, 1970, in order to implement the various requirements included in these Acts. The **Clean Air Act (CAA)** is the comprehensive federal law that regulates air emissions from stationary and mobile sources. Among other things, this law authorizes EPA to establish **National Ambient Air Quality Standards (NAAQS)** to protect public health and public welfare and to regulate emissions of hazardous air pollutants.

One of the goals of the Act was to set and achieve NAAQS in every state by 1975 in order to address the public health and welfare risks posed by certain widespread air pollutants. The setting of these pollutant standards was coupled with directing the states to develop **state implementation plans (SIPs),** applicable to appropriate industrial sources in the state, in order to achieve these standards. The Act was amended in 1977 and 1990 primarily to set new goals (dates) for achieving attainment of NAAQS since many areas of the country had failed to meet the deadlines.

Section 112 of the **Clean Air Act** addresses emissions of hazardous air pollutants. Prior to 1990, CAA established a risk-based program under which only a few standards were developed. The 1990 **Clean Air Act Amendments** revised Section 112 to first require issuance of technology-based standards for major sources and certain area sources. "Major sources" are defined as a stationary source or group of stationary sources that emit or have the potential to emit 10 tons per year or more of a hazardous air pollutant or 25 tons per year or more of a combination of hazardous air pollutants. An "area source" is any stationary source that is not a major source.

For major sources, Section 112 requires that EPA establish emission standards that require the maximum degree of reduction in emissions of hazardous air pollutants. These emission standards are commonly referred to as "maximum achievable control technology" or "MACT" standards. Eight years after the technology-based MACT standards are issued for a source category, EPA is required to review those standards to determine whether any residual risk exists for that source category and, if necessary, revise the standards to address such risk.

U.S. Patriot Act

- Public Law 107-56
- Provides tools required to intercept and obstruct terrorism
- Title 19 Section 2332a

Congress enacted the Patriot Act by overwhelming, bipartisan margins, arming law enforcement with new tools to detect and prevent terrorism: The U.S. Patriot Act was passed nearly unanimously by the Senate 98-1, and 357-66 in the House, with the support of members from across the political spectrum. The Act improves our counterterrorism efforts in several significant ways:

1. **The Patriot Act allows investigators to use the tools that were already available to investigate organized crime and drug trafficking.**

   Many of the tools the Act provides to law enforcement to fight terrorism have been used for decades to fight organized crime and drug dealers, and have been reviewed and approved by the courts. As Senator Joe Biden (D-DE) explained during the floor debate about the Act, "the FBI could get a wiretap to investigate the mafia, but they could not get one to investigate terrorists. To put it bluntly, that was crazy! What's good for the mob should be good for terrorists" (Cong. Rec., 10/25/01).

   - It allows law enforcement to use surveillance against more crimes of terror. Before the Patriot Act, courts could permit law enforcement to conduct electronic surveillance to investigate many ordinary, non-terrorism crimes, such as drug crimes, mail fraud and passport fraud. Agents also could obtain wiretaps to investigate some, but not all, of the crimes that terrorists often commit. The Act enabled investigators to gather information when looking into the full range of terrorism-related crimes, including chemical weapons offenses, the use of weapons of mass destruction, killing Americans abroad and terrorism financing.
   - It allows federal agents to follow sophisticated terrorists trained to evade detection. For years, law enforcement has been able to use "roving wiretaps" to investigate ordinary crimes, including drug offenses and racketeering. A roving wiretap can be authorized by a federal judge to apply to a particular suspect, rather than a particular phone or communications device. Because

international terrorists are sophisticated and trained to thwart surveillance by rapidly changing locations and communication devices such as cell phones, the Act authorized agents to seek court permission to use the same techniques in national security investigations to track terrorists

- It allows law enforcement to conduct investigations without tipping off terrorists. In some cases, if criminals are tipped off too early to an investigation, they might flee, destroy evidence, intimidate or kill witnesses, cut off contact with associates or take other action to evade arrest. Therefore, federal courts in narrow circumstances long have allowed law enforcement to delay for a limited time when the subject is told that a judicially approved search warrant has been executed. Notice is always provided, but the reasonable delay gives law enforcement time to identify the criminal's associates, eliminate immediate threats to our communities and coordinate the arrests of multiple individuals without tipping them off beforehand. These delayed notification search warrants have been used for decades, have proven crucial in drug and organized crime cases and have been upheld by courts as fully constitutional.

- It allows federal agents to ask a court for an order to obtain business records in national security terrorism cases. Examining business records often provides the key that investigators are looking for to solve a wide range of crimes. Investigators might seek select records from hardware stores or chemical plants, for example, to find out who bought materials to make a bomb, or bank records to see who's sending money to terrorists. Law enforcement authorities have always been able to obtain business records in criminal cases through grand jury subpoenas and continue to do so in national security cases where appropriate. These records were sought in criminal cases such as the investigation of the Zodiac gunman, where police suspected the gunman was inspired by a Scottish occult poet, and wanted to learn who had checked the poet's books out of the library. In national security cases where use of the grand jury process was not appropriate, investigators previously had limited tools at their disposal to obtain certain business records. Under the Patriot Act, the government can now ask a federal court (the Foreign Intelligence Surveillance Court), if needed to aid an investigation, to order production of the same type of records available through grand jury subpoenas. This federal court, however, can issue these orders only after the government demonstrates the records concerned are sought for an authorized investigation to obtain foreign intelligence information not concerning a U.S. person or to protect against international terrorism or clandestine intelligence activities, provided that such investigation of a U.S. person is not conducted solely on the basis of activities protected by the First Amendment.

2. The Patriot Act facilitated information sharing and cooperation among government agencies so that they can better "connect the dots."

The Act removed the major legal barriers that prevented the law enforcement, intelligence and national defense communities from talking and coordinating their work to protect the American people and our national security. The government's prevention efforts should not be restricted by boxes on an organizational chart. Now police officers, FBI agents, federal prosecutors and intelligence officials can protect our communities by "connecting the dots" to uncover terrorist plots before they are completed. As Senator John Edwards (D-NC) said about the Patriot Act, "we simply cannot prevail in the battle against terrorism if the right hand of our government has no idea what the left hand is doing" (Press release, 10/26/01).

- Prosecutors and investigators used information shared pursuant to Section 218 in investigating the defendants in the so-called Virginia Jihad case. This prosecution involved members of the Dar al-Arqam Islamic Center, who trained for jihad in Northern Virginia by participating in paintball and paramilitary training, including eight individuals who traveled to terrorist training camps in Pakistan or Afghanistan between 1999 and 2001. These individuals are associates of a violent Islamic extremist group known as Lashkar-e-Taiba (LET), which operates in Pakistan and Kashmir, and that has ties to the al-Qaeda terrorist network. As the result of an investigation that included the use of information obtained through Foreign Intelligence Surveillance Act of 1978 (FISA), prosecutors were able to bring charges against these individuals. Six of the defendants have pleaded guilty, and three were convicted in March 2004 of charges including conspiracy to levy war against the United States and conspiracy to provide material support to the Taliban. These nine defendants received sentences ranging from a prison term of 4 years to life imprisonment.

3. The Patriot Act updated the law to reflect new technologies and new threats.

The Act brought the law up to date with current technology, so we no longer have to fight a digital-age battle with antique weapons-legal authorities leftover from the era of rotary telephones. When investigating the murder of *Wall Street Journal* reporter Daniel Pearl, for example, law enforcement used one of the Act's new authorities to use high-technology means to identify and locate some of the killers.

- It allows law enforcement officials to obtain a search warrant anywhere a terrorist-related activity occurred. Before the Patriot Act, law enforcement personnel were required to obtain a search warrant in the district where they intended to conduct a search. However, modern terrorism investigations often span a number of districts, and officers therefore had to obtain multiple warrants in multiple jurisdictions, creating unnecessary delays. The Act provides that warrants can be obtained in any district in which terrorism-related activities occurred, regardless of where they will be executed. This provision does not change the standards governing the availability of a search warrant, but streamlines the search warrant process.
- It allows victims of computer hacking to request law enforcement assistance in monitoring the "trespassers" on their computers. This change made the law technology neutral; it placed electronic trespassers on the same footing as physical trespassers. Now, hacking victims can seek law enforcement assistance to combat hackers, just as burglary victims have been able to invite officers into their homes to catch burglars.

4. **The Patriot Act increased the penalties for those who commit terrorist crimes.**

   Americans are threatened as much by the terrorist who pays for a bomb as by the one who pushes the button. That's why the Patriot Act imposed tough new penalties on those who commit and support terrorist operations, both at home and abroad. In particular, the Act

   - Prohibits the harboring of terrorists. **The Act** created a new offense that prohibits knowingly harboring persons who have committed or are about to commit a variety of terrorist offenses, such as destruction of aircraft; use of nuclear, chemical or biological weapons; use of weapons of mass destruction; bombing of government property; sabotage of nuclear facilities and aircraft piracy.
   - Enhanced the inadequate maximum penalties for various crimes likely to be committed by terrorists, including arson, destruction of energy facilities, material support to terrorists and terrorist organizations and destruction of national-defense materials.
   - Enhanced a number of conspiracy penalties, including for arson, killings in federal facilities, attacking communications systems,

material support to terrorists, sabotage of nuclear facilities and interference with flight crew members. Under previous law, many terrorism statutes did not specifically prohibit engaging in conspiracies to commit the underlying offenses. In such cases, the government could only bring prosecutions under the general federal conspiracy provision, which carries a maximum penalty of **only five** years in prison.
- Punishes terrorist attacks on mass transit systems.
- Punishes bioterrorists.
- Eliminates the statutes of limitations for certain terrorism crimes and lengthens them for other terrorist crimes.

The government's success in preventing another catastrophic attack on the American homeland since September 11, 2001, would have been much more difficult, if not impossible, without the U.S.A Patriot Act. The authorities Congress provided have substantially enhanced our ability to prevent, investigate and prosecute acts of terror.

Resource Conservation and Recovery Act (RCRA) Laws and Regulations (RCRA)

- Mandated hazardous waste trucking
- Cradle-to-grave power
- Toxic waste disposal facility
- Facility licensing and permitting

What Is RCRA?

The term Resource Conservation and Recovery Act (RCRA) is often used interchangeably to refer to the law, regulations and EPA policy and guidance. The law describes the waste management program mandated by Congress that gave EPA authority to develop the RCRA program. EPA regulations carry out the congressional intent by providing explicit, legally enforceable requirements for waste management. These regulations can be found in title 40 of the **Code of Federal Regulations (CFR),** parts 239 through 282. EPA guidance documents and policy directives clarify issues related to the implementation of the regulations. The components that make up the RCRA program include the Act, Regulations, Policy and Guidance.

How Does RCRA Work?

RCRA establishes the framework for a national system of solid waste control. Subtitle D of the Act is dedicated to nonhazardous solid waste requirements, and Subtitle C focuses on hazardous solid waste. Solid waste includes solids, liquids and gases, and must be discarded to be considered waste. Congress has amended RCRA several times, which requires the President's signature to become law. EPA translates this direction into operating programs by developing regulations, guidance and policy. States play the lead role in implementing nonhazardous waste programs under Subtitle D. EPA has developed regulations to set minimum national technical standards for how disposal facilities should be designed and operated. States issue permits to ensure compliance with EPA and state regulations. The regulated community is comprised of a large, diverse group that must understand and comply with RCRA regulations. These groups can include hazardous waste generators, government agencies and small businesses and gas stations with underground petroleum tanks.

Subtitle C—Hazardous Waste

Hazardous waste is regulated under Subtitle C of RCRA. EPA has developed a comprehensive program to ensure that hazardous waste is managed safely from the moment it is generated to its final disposal (cradle-to-grave). Under Subtitle C, EPA may authorize states to implement key provisions of hazardous waste requirements in lieu of the federal government. If a state program does not exist, EPA directly implements the hazardous waste requirements in that state. Subtitle C regulations set criteria for hazardous waste generators, transporters, and treatment, storage and disposal facilities. This includes permitting requirements, enforcement and corrective action or cleanup.

Subtitle D—Nonhazardous Waste

Nonhazardous solid waste is regulated under Subtitle D of RCRA. Regulations established under Subtitle D ban open dumping of waste and set minimum federal criteria for the operation of municipal waste and industrial waste landfills, including design criteria, location restrictions, financial assurance, corrective action (cleanup) and closure requirement. States play a lead role in implementing these regulations and may set more stringent requirements. In absence of an approved state program, the federal requirements must be met by waste facilities.

RCRA Today

EPA has largely focused on building the hazardous and municipal solid waste programs, and fostering a strong societal commitment to recycling and pollution prevention. Ensuring responsible waste management practices is a far-reaching and challenging task that engages EPA headquarters, regions, state agencies, tribes and local governments, as well as everyone who generates waste. It is important to look at the RCRA program's nationwide accomplishments to understand where it is now and where it is headed in the future.

- Developing a comprehensive system and federal/state infrastructure to manage hazardous waste from "cradle to grave"
- Establishing the framework for states to implement effective municipal solid waste and nonhazardous secondary material management programs
- Preventing contamination from adversely impacting our communities and resulting in future Superfund sites
- Restoring 18 million acres of contaminated lands, nearly equal to the size of South Carolina, and making the land ready for productive reuse through the RCRA Corrective Action program
- Creating partnership and award programs to encourage companies to modify manufacturing practices in which to generate less waste and reuse materials safely
- Enhancing perceptions of wastes as valuable commodities that can be part of new products through its sustainable materials management efforts
- Bolstering the nation's recycling infrastructure and increasing the municipal solid waste (MSW) recycling/composting rate from less than **seven percent to about 34.6 percent.**

The RCRA program has evolved in response to changes in waste generation and management aspects that could not have been foreseen when the program was first put in place. The RCRA program is needed to address continuing challenges, including the following:

- Highly toxic waste
- Wastes from increasingly efficient air and water pollution control devices
- Population growth that places larger demands on our natural resources
- Long-term stewardship of facilities that closed with waste in place
- Looking toward the future, it is important for the RCRA program to continue to fulfill its mission by
  - Continuing to safeguard communities and the environment.
  - Mitigating and cleaning up contamination.
  - Championing sustainable, lifecycle waste and material management approaches.
  - Promoting economic development (including job creation) and community well-being.
  - Embracing technological advances that will facilitate commerce and enhance stakeholders' participation in the decisions affecting their communities.

Federal Regulations

Federal regulations and the Code of Federal Regulations CFR.
- Fifty sections known as titles.
  a. Not law, but authorized by law.
  b. Developed (promulgated) by governmental authorities.
  c. Follows standard process of publishing and comment periods. The standard process involves publication in the "Federal Register of Intended Rule Making." This is opened for a comment period. The authority then publishes interim final rules and opens a comment period and possibly public hearings. The authority evaluates comments and then publishes final rules.
  d. When final, they carry the weight of law.
  e. They are mandatory standards for those falling within the scope and application of the regulation.

40 CFR—EPA

  a. 40 CFR 300 National Oil and Hazardous Substance Contingency Plan
  b. Also known as the NCP
  c. Mechanism for handling chemical emergencies
  d. Includes NRT and all of its requirements
  - **National Response Center (NRC).**
  - **Regional Response Teams (RRTs).**
  - **Federal On Scene Coordinators (OSC).**

40 CFR 68—Risk Management Planning

  a. Identifies substances and quantities subject to the requirement
  b. Offsite consequence scenario analysis
  - Worst-case
  - Alternative
- Documentation of 5-year accident history for covered process
- Integrated prevention program
- Emergency response program to address identified needs
- Identify overall management for implementation
- Must be revised every 5 years
- Sites with limited potential to affect off-site locations (program 1) need not meet the requirements listed under d through f above
- Site with potential to affect off-site locations (program 2 and 3 facilities) must meet all requirements

40 CFR 310—Emergency Response Financial Restitution (Figure 2.28)

- Up to $25,000 for restitution of response costs that cannot be collected through any other means
- Full documentation of costs is required
- Only eligible for costs incurred after notifying the NRC
- Must have functioning LEPC and LEPC plans

40 CFR 311—EPA HAZWOPER Regulations

a. EPA's version of 1910.120
b. Defines employee
c. Refers to 29 CFR 1910.120

355 et al.-EPCRA Planning

a. Regulatory requirements of SARA Title III
b. Now routinely called EPCRA

Standard of Care

EPA 311 and OSHA 1910.120

Federal regulations and standards that form a mandatory standard of care that first responders must follow.

Figure 2.28 Emergency Response Financial Restitution: Up to $25,000 for restitution of response costs that cannot be collected through any other means.

NFPA standards

Not mandatory unless adopted by authority having jurisdiction.

Organizational standard operating procedures (SOPs) and standard operating guidelines (SOGs).

29 CFR 1910.120

    a. Final rule became effective in March 1989
    b. Contains 17 paragraphs
    c. Has five appendices
    d. Particular importance to **fire service.**
      • Medical surveillance
      • Emergency response program to hazardous materials substance release

Elements/paragraphs of 29 CFR 1910.120: scope, application, and definitions.

    a. Scope: This section covers the following operations, unless the employer can demonstrate that the operation does not involve employee exposure or the reasonable possibility for employee exposure to safety or health hazards.
    b. Sections (a)(1)(i) through (a)(1)(v) describe those who are covered by this document.
      • Emergency response operations for releases of, or substantial threats of releases of, hazardous substances without regard to the location of the hazard ((a)(1)(v)).
        a. Safety and health program (b)
        b. Site characterization and analysis (c)
        c. Site control (d)
        d. Training (e)
        e. Medical surveillance (f)
      • General: Employees engaged in operations specified in paragraphs (a)(1)(i) through (a)(1)(1v) of this section and not covered by (a)(2)(iii) exceptions and employees of employers specified in paragraph (q)(9) shall institute a medical surveillance program in accordance with this paragraph.
        Members of an organized and designated hazmat team and hazardous materials specialists shall receive a baseline physical examination and be provided with medical surveillance as required in paragraph (f) of this section.
    a. Engineering controls, work practices and personal protection equipment for employees' protection (g)

b. Monitoring (h)
c. Handling drums and containers (i)
d. Opening drums and containers (j)
e. Decontamination (k)
f. Emergency response by employees at an uncontrolled hazardous waste site (I). Note: this is for employees working at the hazardous waste site
g. Illumination (m)
h. Non-potable water (n)
i. New technology programs (o)
j. Certain operations conducted under the **Resource Conversation and Recovery Act of 1976 (RCRA)** (p)
k. Emergency response program to hazardous substance release (q)
   This paragraph is intended for emergency response personnel who, as part of their normal assignments, may become involved with a released hazardous substance such as EMS, fire, law enforcement, public works, sanitation workers, and others.

29 CFR 1910.120 Appendices

a. Personal protective equipment test methods (A)
b. General description and discussion of the levels of protective gear (B)
c. Compliance guidelines (C)
d. References (D)
e. Training curriculum guidelines (nonmandatory) (E)

Interpretations

a. While the standard has not been amended since 1989, its interpretation changes on a regular basis.
b. Interpretations include items such as
   • What level of training certain responders must receive.
   • What operations can be performed at the Operations Level.

Appendix E: Training Curriculum Guidance

a. The following nonmandatory general criteria may be used for assistance in developing site-specific training curriculum to meet the training requirements of 29 CFR 1910.120 including (q)(6), (q)(7) and (q)(8).
b. Note: (q)(6) through (q)(8) are components of the Emergency Response Program to Hazardous Substance Release chapter—first responders.

According to OSHA 29 CFR 1910.120(q), every emergency responder who responds to the scene of a hazardous materials incident must be trained to a minimum of the awareness level. Over the years, interpretations have seemed to have clarified the awareness level, and it appears that awareness applies to employees that might come across a hazardous materials incident during their daily job functions (Figure 2.29). They need to know what to do based upon awareness level training. This requirement has been around since the mid-1980s, and to date, there are employees who do not have hazardous materials awareness training. That is the equivalent to sending firefighters to a fire without basic training, EMS personnel to medical emergencies without medical training and police officers to an armed robbery without firearms training.

OSHA regulations aside, it just makes good sense to prepare emergency responders to deal with hazardous materials. Responding to acts of terrorism has also become the target of specialized awareness training for emergency responders. I mention terrorism in a hazardous materials book, because I believe terrorism is a form of hazardous materials incident, with special circumstances. Response to terrorism requires training for first responders in addition to awareness training required to respond to hazardous materials incidents. Chief John Eversole of the Chicago Fire Department, a hazmat pioneer, has always said, "if you cannot do Hazmat, you can't do terrorism." There is a lot of truth in that statement. Therefore, we will address terrorism awareness competencies in this volume in addition to hazardous materials awareness training.

Figure 2.29 Over the years, interpretations have seemed to have clarified the awareness level, and it appears that awareness applies to employees that might come across a hazardous materials incident during their daily job functions.

When the OSHA requirements first surfaced, it was difficult to convince the response community the training was needed. There were chiefs who declared, "we won't respond to hazardous materials incidents then we won't require the training." Well think about it, how many hazardous materials incidents or other incidents that have the potential for hazardous materials are called in or dispatched as hazardous materials incidents? Many emergencies are fires, EMS responses, traffic accidents or crime events, which involve hazardous materials as an additional problem. Many times responders are not aware of what to look for in a potential hazardous materials situation. They may become victims of the hazardous materials and not know it until they experience the injury, illness or death from the exposure.

OSHA regulations do not identify an amount of time responders must be exposed to awareness training, but rather identify certain knowledge areas in which the responders must prove proficiency. Response personnel, including fire, EMS, police, public works and certain others, should have a minimum of awareness level training.

Standards

NFPA 472—Standard for Competence of Responders to Hazardous Materials/Weapons of Mass Destruction Incidents (Figure 2.30)

- List of competencies for response at various levels of capability and incident scene responsibility
- Revised and updated on a regular basis
- Fifteen chapters and eight Annex chapters
  Chapter 1: Administration
  Chapter 2: Reference Publications

Figure 2.30 NFPA 472—Standard for Competence of Responders to Hazardous Materials/Weapons of Mass Destruction Incidents: List of competencies for response at various levels of capability and incident scene responsibility.

Chapter 3: Definitions
Chapter 4: Competencies for Awareness Level Personnel
Chapter 5: Competencies for Operations Level Personnel
Chapter 6: Mission Specific for Operations Level Personnel
    6.1: Introduction
    6.2: Personal protective equipment
    6.3: Technical and mass decontamination
    6.4: Evidence preservation and sampling
    6.5: Product control
    6.6: Air monitoring and sampling
Chapter 7: Competencies for Hazardous Materials Technician
Chapter 8: Competencies for Incident Commanders
    8.1: General
    8.2: Analyzing the incident
    8.3: Planning the response
    8.4: Implementing the response
    8.5: Evaluating progress
    8.6: Terminating the incident
Chapter 9: Competencies for Specialist Employee
Chapter 10: Competencies for Hazardous Materials Officer (geared toward WMD-related incidents)
Chapter 11: Competencies for Hazardous Materials Safety Officer
Chapter 12: Competencies for Hazardous Materials Technician with a Tank Car Specialty
Chapter 13: Competencies for Hazardous Materials Technician with a Cargo Tank Specialty
Chapter 14: Competencies for a Hazardous Materials Technician with an Intermodal Tank Specialty
Chapter 15: Competencies for a Hazardous Materials Technician with a Marine Tank/Vessel Specialty

NFPA 472—Annex materials

- Explanatory material (A)
- Competencies for a technician with a flammable liquid bulk storage tank specialty (B)
- Competencies for a technician with a flammable gas bulk storage specialty (C)
- Competencies for a technician with a radioactive material specialty (D)
- Overview of responder levels and tasks at hazardous materials/WMD incidents (E)
- Definitions of hazardous materials (F)
- United Nations (UN)/DOT hazard classes and divisions (G)
- Informational resources (H)

NFPA 473—Standard for Competencies for EMS Personnel Responding to Hazardous Materials and Weapons of Mass/Destruction Incidents

Chapter 4: Basic Life Support Competencies
Chapter 5: Regulations versus Standards
- Developed by individuals in the field
- Nonmandatory unless adopted
- Contribute to the standard of care
- Help establish the "reasonable person"

Agencies

Department of Energy, DOT, Nuclear Regulatory Commission and EPA

Radioactive Materials Several agencies have overlapping authorities for regulating shipments of radioactive materials (Figure 2.30A). DOT regulates the shipment of hazardous materials, including radioactive materials. The Nuclear Regulatory Commission (NRC) regulates commercial activities of nuclear power plants. Department of Energy (DOE) ships commercial radioactive waste for storage and defense nuclear waste and weapons for storage or use. DOE and EPA share responsibility for transportation of hazardous wastes or radioactive and hazardous waste mixtures generated at facilities operated by DOE under the authority of the Atomic Energy Agency(**AEA**).

Figure 2.30A Several agencies have overlapping authorities for regulating shipments of radioactive materials.

Government Resources

National Transportation Safety Board (NTSB) Every emergency responder should be familiar with the National Transportation Safety Board (NTSB) the agency charged with the investigation of transportation incidents in this country to increase safety in all modes of transportation. Whenever an accident involving aircraft, trains or buses occurs, NTSB is quick to mobilize and arrive on the scene to start investigating the cause. **They** were heavily involved in the investigation of TWA Flight 800 along with the FBI and other agencies. However, their jurisdiction is limited to transportation accidents. The NTSB's jurisdiction, however, is limited to transportation accidents.

United States Chemical Safety and Hazard Investigation Board

Chemical Safety Board (CSB) 20 Years (1998–2018) During 2018, the Chemical Safety Board (CSB) celebrated its 20th year in operation (Figure 2.31). In spite of President Donald Trump's intent to end the agency as part of a government downsizing, CSB has performed an outstanding service to the chemical industry and the emergency response community. Congress has continued to fund the agency, and they have been conducting investigations throughout 2018. Today the agency has three board members: Interim Executive authority is Kristen Kulinowski and assisting her are two board members, Manuel "Manny" Ehrlich and Rick Engler. Hopefully Congress and the President will decide to keep funding this import government agency.

When a propane railcar exploded in Kingman, AZ, on July 5, 1973, killing 11 firefighters and one civilian, the NTSB responded but could not investigate because the tank was on a rail siding and not in transit. Emergency officials repeatedly urged the creation of a similar federal agency to investigate accidents involving chemicals in transportation, storage and use.

Chemicals have the potential to cause a catastrophe, not only to industry workers but to surrounding communities. Taking a proactive approach can have a positive outcome in preventing these types of

Figure 2.31 During 2018, the **Chemical Safety Board** celebrated its 20th year in operation.

disasters before they occur. In 1989, Congress searched the federal system to find any existing federal resources to deal with chemical accidents. A Senate report concluded, "Based on ... review of the record with respect to chemical accidents and the measures that various federal agencies, states and other nations are taking to respond to chemical accidents, several conclusions are warranted. Accident prevention has great promise, but is not given sufficient attention in current federal programs. No agency of the United States government is actively engaged in efforts to prevent chemical accidents, today." Fourteen federal agencies were actively involved in accident prevention at that time, so Congress further identified a need to "... improve the effectiveness of accident prevention programs and reduce the burden of duplicative requirements of regulated entities." Industry representatives voiced support for the concepts in congressional hearings.

After evaluating alternatives, Congress concluded there was a need to determine the causes of chemical accidents and identify solutions for prevention of future incidents. Congress determined that the best way to accomplish this task would be through the establishment of a national chemical safety and hazard investigation board. Thus, the CSB was authorized. The first steps were taken in 1990, when Congress attached amendments to Public Law 101-549, the **Clean Air Act**, authorizing the creation of an independent federal agency to concentrate on the investigation of chemical accidents. The U.S. Chemical Safety and Hazard Investigation Board (CSB) was born.

After several years of growing pains as well as attempts by the Clinton Administration to eliminate it through budget cuts, the CSB began operations on January 5, 1998. It had already been involved in more than 15 chemical accidents, including a propane explosion in Iowa that killed two volunteer firefighters. Since that time, the CSB has investigated over 130 chemical incidents and made over 800 recommendations based upon the investigations. These accidents have resulted in over 144 deaths and at least 780 serious injuries. Incidents investigated have occurred in 37 states: Alabama (2), Arizona, Arkansas, California (5), Colorado, Connecticut, Delaware (3), Florida (3), Georgia (3), Hawaii, Idaho, Illinois (3), Indiana (2), Iowa (2), Kansas (2), Kentucky (4), Louisiana (7), Massachusetts, Mississippi (4), Missouri (3), Nevada (2), New Jersey (3), New Mexico, New York (2), North Carolina (4), Ohio (4), Oklahoma (2), Oregon, Pennsylvania (3), Rhode Island, Tennessee, Texas (14), Utah, Virginia, West Virginia (6), Wisconsin (3), and Washington State. One incident occurred off shore in the Gulf, one in Washington DC and one in Puerto Rico.

The board's mission is to "work in concert with industry, labor, government and communities to help prevent accidents" involving chemicals. This mission is to be accomplished by

- Investigating and issuing reports on causes of chemical accidents.
- Assessing the effectiveness of federal agencies in preventing chemical accidents.
- Eliminating duplication of efforts at the federal level concerning chemical accident investigation and prevention.
- Conducting special studies.
- Providing the findings of investigations and research to industry to improve safety of operations.

Further, the board has been charged by Congress with providing an independent evaluation of chemical accident causes, advancing approaches for mitigating the problems and describing means to consolidate and curtail costs of the government's accident investigation operations. Congress has further mandated that the board investigate or cause to be investigated any and all accidental releases of any toxic or hazardous chemical resulting in a fatality, serious injury or substantial property damage.

Initial Operations

During the board's first year of operation, due to limited funding, its staff included five board members, one senior advisor, four special assistants, five technical professionals and four secretaries/administrative assistants. The board was headed initially by Paul Hill Jr, PhD. Hill brought many years of chemical experience to the board, having headed the National Institute of Chemical Studies (NICS), which was a nonprofit organization with a mission of public education, emergency preparedness, community safety, pollution prevention, hazard identification, risk reduction and risk communication concerning toxic and hazardous chemicals.

Gerald V. Poje, PhD, a board member, specialized in policy for toxicology and chemical hazards. His background includes directing the National Institute of Environmental Health Sciences. Phyllis G. Thompson, PhD, is a special policy assistant for the board.

Author's Note: I had the pleasure of knowing her and worked with Thompson during her tenure with the Federal Emergency Management Agency (FEMA) as the technical training officer for the Chemical Stockpile Emergency Preparedness Program (CSEPP), which dealt with training issues for the Army's disposal of U.S. stockpiles of chemical weapons. During that time, I was impressed with her knowledge of training and chemical safety issues along with her dedication to FEMA and the CSEPP program.

Christopher W. Warner was a special assistant for legal operations. Warner, a lawyer, is a former advisor to the Interior Department. He will be the general counsel for the board. Phillip S. Cogan was a special assistant for external relations. He is a former print and broadcast journalist and was the lead public affairs officer for 88 presidential disaster declarations. The office of external affairs provides intergovernmental relations, acts as a liaison with Congress, prepares publications and is the source of public information concerning the board's activities.

How to Report Incidents

All chemical accidents resulting in deaths, serious injuries or significant property damage should be reported to the CSB. If such a chemical release occurs, the **National Response Center (NRC)** must be notified at its toll-free number (800-424-8802). The center is operated 24 hours a day, 365 days a year, by the **U.S. Coast Guard**. In addition to being the contact point for the **Chemical Safety and Hazard Investigation Board**, the NRC receives reports of other types of chemical releases required under various federal laws and regulations.

Investigations Begin

The paint was hardly dry on the CSB's door signs when it was plunged into its first investigation on January 7, 1998. An explosion at the Sierra Chemical Company near Reno, NV, killed four workers and injured three others. Sierra Chemical manufactures explosives for mining operations.

Two explosions occurred in a building used to manufacture boosters. Falling debris from those blasts triggered a third explosion involving an explosives-storage building at the south end of the company site. The cause of the initial blasts is believed to be from a worker leaving materials in a mixing pot overnight. The following morning, another worker turned on the pot and its blade hit some solidified explosive material, triggering an electric shock wave that initiated the explosion. Board findings of circumstances that attributed to the tragedy included a lack of worker training, poor regulatory oversight and deficiencies in safety procedures at the facility. The final report is still pending at this time. The sequence of events that takes place when the CSB is called upon to investigate an accident includes sending personnel to the scene to interview people and collect physical evidence, holding a board of inquiry, holding public hearings and issuing a final report.

On March 4, 1998, four workers were killed at Sonat Exploration, an oil-separation facility near Pitkin, LA. Another incident occurred on March 27, 1998, at the Quest Aerospace facility near Yuma, AZ, where toy rocket motors are manufactured. The explosion killed four workers.

On the same day, a confined space accident occurred at a Union Carbide plant in Hahnville, LA. One worker was killed and another seriously injured. Nine workers were injured during a fire on April 8, 1998, at the Morton Specialty Chemical Company in Paterson, NJ.

Albert City, IA, April 8, 1990, Propane Explosion

Two volunteer firefighters were killed on April 8, 1998, in a propane explosion on a poultry farm in Albert City, IA, northwest of the capital city of Des Moines. Seven others, including a deputy sheriff, were injured in the explosion. A pipe connected to a large propane storage tank was broken by children riding four-wheel all-terrain vehicles. The leaking propane found an ignition source and ignited. Eventually, a BLEVE (boiling liquid expanding vapor explosion) occurred in the propane tank as a result of direct flame impingement from the burning propane.

In addition to the investigation by the CSB, the National Institutes for Occupational Safety and Health (NIOSH) investigated the accident. NIOSH has been directed by Congress to investigate all firefighter line-of-duty deaths and develop recommendations to prevent future injuries and deaths. Current and completed investigations can be viewed and downloaded from CSB's website.

In early September 1998, an incident occurred involving liquid nitrogen, a cryogenic liquid that presents an asphyxiation hazard in addition to being very cold. The accident occurred while two workers were working on an oil pipeline in Springer, OK. They had been using liquid nitrogen in a pit to pressure check the pipeline and (in the opinion of the author) were likely to have been asphyxiated by the nitrogen displacing the oxygen in the pit. When discovered, both men were frozen by the very cold nitrogen, which has a liquid boiling point of 320° below zero Fahrenheit; the vapors being released by the liquid would also have been very cold.

United States Department of Transportation (DOT)

Regulations

Responder Assistance Emergency Response Guidebook (First Responders' "Bible") The **United States Department of Transportation (DOT)**, creator and distributor of the Emergency Response Guidebook (ERG) celebrated its 50th year since it was established in 2016. They released the 2020 edition of the ERG in the late summer of 2020. More than 1.5 million free copies have been provided to first responders nationwide by the DOT through state contacts in each state. During 2019 I spoke with the DOT Pipeline and Hazardous Materials Safety Administration (PHMSA) about the latest edition of the ERG. My main focus was "What in particular would they want to have first responders know about the ERG?" Their response was "emphasize the fact that the ERG is intended for use by first responders during the initial phase (first 30 minutes) of a transportation incident involving hazardous materials and is not intended for use during incidents at fixed facilities." Additionally, DOT also wanted responders to **RESIST RUSHING IN!** Hazardous materials incidents do not occur often in many of the jurisdictions throughout the United States, Canada and Mexico. They are technical in nature and can be extremely dangerous to response personnel. It is important that first responders exercise competencies outlined by OSHA 1910.120 and NFPA Standards 472-473, which are designed to keep them safe at the scene of a hazardous materials incident while dictating their limitations in terms of training and equipment.

Evolution of the Emergency Response Guidebook
The field of hazardous materials is a relatively new subject for emergency responders. DOT developed the first version of a responder guide book in 1977 (Figure 2.32). This new responder resource was called the Hazardous Materials Action Guides, which was 87 pages long and contained information on 43 chemicals. During 1980, the DOT created the first version of the ERG utilizing the format we are familiar with today. It contained 66 orange guides, placard chart and numerical and alphabetical sections although they were not yet color coded as they are in later versions. There have been twelve editions of the ERG to date: 1980, 1984, 1987, 1990, 1993, 1996, 2000, 2004, 2008, 2012, 2016 and 2020 (Figure 2.33). The next version is due out in 2024. By comparison, the 2020 version of the ERG contains 400 pages and lists hundreds of chemicals. Because of the quantity of information available to first responders for dealing with the initial stages of a hazardous materials or dangerous goods incident responders must receive training on its use for it to be o10f greatest benefit at an incident.

> **Author's Note:** *Over the years, I have managed to collect every version of the ERG from the very first one. That is where I got the photograph of the covers of all the editions.*

Emergency Response Guidebook
Each issue of the ERG is on a 4-year cycle, and work begins on the next edition soon after the current one is published. The ERG working group consists of government representatives from the United States DOT/PHMSA, Transport Canada and Secretariat of Transport and Communications in Mexico along with a representative from Argentina's emergency response call center, CEQUIME. Comments were solicited from ERG users and stakeholders through an announcement in the *Federal*

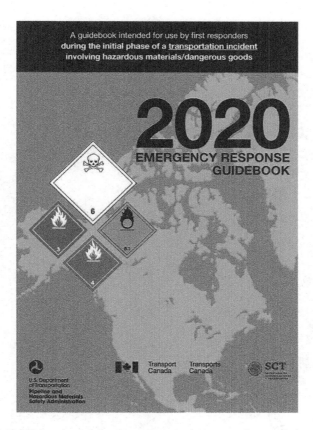

Figure 2.32 DOT developed the first version of a responder guide book in 1977.

Register in September 2014, an announcement on Transport Canada's website in August 2014, and through face-to-face outreach engagements at hazmat conferences throughout the country including the IAFC Response Teams Conference, Baltimore, MD (May 2014); Continuing Challenge, Sacramento, CA (September 2014); and HOTZONE in Houston, TX (October 2014).

Now located on page 1 of the 2020 ERG is How to Use This Guidebook Flow Chart (Figure 2.34). At the top are instructions for approach to the incident scene from upwind, uphill or upstream. Responders are also instructed to stay clear of all spills, vapors, fumes, smoke and potential hazards. The following **warning** is also shown at the top of the flowchart **"Do not use the flowchart if more than one hazardous material/dangerous good is involved. Immediately call the appropriate emergency response agency telephone number listed on the inside back cove of**

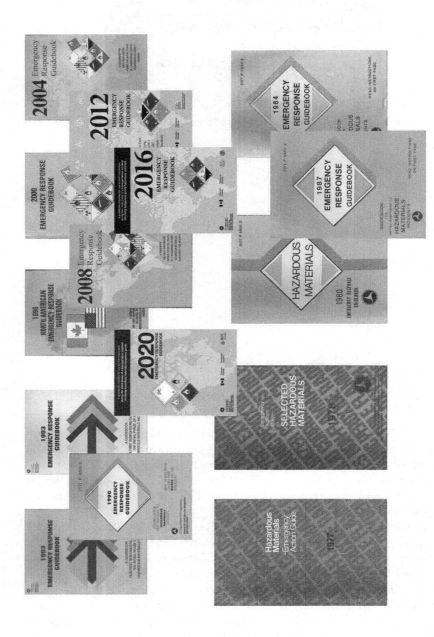

Figure 2.33 There have been nine editions of the ERG to date: 1980, 1984, 1987, 1990, 1993, 1996, 2000, 2004, 2008, 2012 and 2016.

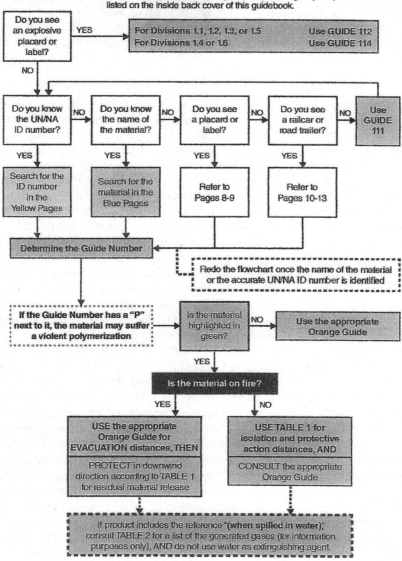

Figure 2.34 Now located on page 1 of the 2020 ERG is a How to Use This Guidebook Flow Chart.

this guidebook." This information is intended to improve the safety of emergency response personnel as they respond to a hazardous materials incident scene. Page 2 of the 2020 ERG is now an innovative and useful page for entering local emergency telephone numbers for local assistance. Page 3 is now Table of Contents to provide for quicker navigation through the book at the incident scene. These items along with the Flowchart are excellent in the 2020 ERG. Following these additions are the usual pages leading into the main body of the book, including safety precautions and notification instructions.

Initial Response Actions

OSHA 1910.120 and NFPA 472 identify several responsibilities for first responders to a hazardous or weapons of mass destruction incident. These include recognition, isolation and protection. The ERG addresses all of these first responders' duties and provides resources to help carry them out. Without being able to recognize when a hazardous materials incident has occurred, nothing else really matters. So, recognition is the most important criteria when responding to any incident scene. That is why it is important to resist rushing in. The ERG has many resources to help in recognition, including the placard charts, railcar charts (Figure 2.34A,B), road trailer charts (Figure 2.34C,D), Globally Harmonized System (GHS) labels (Figure 2.34E) and pipeline section (Figure 2.34F). Once it is determined that a hazardous materials incident has occurred, in general first responders do not have the necessary training or equipment to mitigate the incident. Response personnel need to access the Yellow and Blue Guide sections of the ERG to determine a name for a hazardous material or find a generic Orange Guide page based upon placards, tank car, road trailer or pipeline information in the ERG. Within the ERG are several notification resources including CHEMTREC (Figure 2.35), the **National Response Center**, military shipments, National Poison Control Center, as well as some private response companies. Also in the 2020 ERG is a page you can fill in with local response contact information, including local railroad contacts. Following proper notifications, the incident scene needs to be isolated so that responders and the public do not come in contact with the hazardous material. ERG contains information on initial isolation distances in the Orange and Green Guide sections. Finally, first responders need to protect themselves and the public from harm created by the release of a hazardous material. Protection information for responders is identified in the Orange section of the ERG. Evacuation distances to protect the public are located in the Orange and Green sections. Initial isolation distances and evacuation distances are determined by computer modeling and analysis of actual incidents, and some distances may change from one edition of the ERG to another based on changing information from computer models and actual incidents.

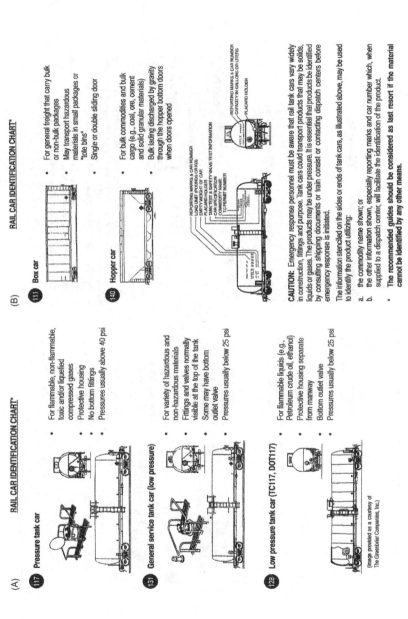

(A)

RAIL CAR IDENTIFICATION CHART*

117 Pressure tank car

- For flammable, non-flammable, toxic and/or liquefied compressed gases
- Protective housing
- No bottom fittings
- Pressures usually above 40 psi

131 General service tank car (low pressure)

- For variety of hazardous and non-hazardous materials
- Fittings and valves normally visible at the top of the tank
- Some may have bottom outlet valve
- Pressures usually below 25 psi

128 Low pressure tank car (TC117, DOT117)

- For flammable liquids (e.g., Petroleum crude oil, ethanol)
- Protective housing separate from manway
- Bottom outlet valve
- Pressures usually below 25 psi

(Image provided as a courtesy of The Greenbrier Companies, Inc.)

(B)

RAIL CAR IDENTIFICATION CHART*

111 Box car

- For general freight that carry bulk or non-bulk packages
- May transport hazardous materials in small packages or "tote bins"
- Single or double sliding door

140 Hopper car

- For bulk commodities and bulk cargo (e.g., coal, ore, cement and solid granular materials)
- Bulk lading discharged by gravity through the hopper bottom doors when doors opened

REPORTING MARKS & CAR NUMBER
LOAD LIMIT (POUNDS OR KG)
EMPTY WEIGHT OF CAR
PLACARD HOLDER
TANK TEST & SAFETY VALVE TEST INFORMATION
CAR SPECIFICATION
COMMODITY NAME
TC PERMIT NUMBER

REPORTING MARKS & CAR NUMBER
CAPACITY IN GALLONS OR LITERS
PLACARD HOLDER

CAUTION: Emergency response personnel must be aware that rail tank cars vary widely in construction, fittings and purpose. Tank cars could transport products that may be solids, liquids or gases. The products may be under pressure. It is essential that products be identified by consulting shipping documents or train consist or contacting dispatch centers before emergency response is initiated.

The information stenciled on the sides or ends of tank cars, as illustrated above, may be used to identify the product utilizing:

a. the commodity name shown; or
b. the other information shown, especially reporting marks and car number which, when supplied to a dispatch center, will facilitate the identification of the product.

* **The recommended guides should be considered as last resort if the material cannot be identified by any other means.**

Figure 2.34A and B Railcar charts.

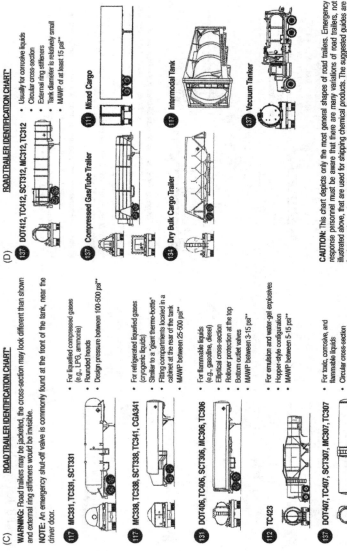

Figure 2.34C and D Road trailer charts.

GHS Pictograms	Physical hazards	GHS Pictograms	Health and Environmental hazards
	Explosive; Self-reactive; Organic peroxide		Skin corrosion; Serious eye damage
	Flammable; Pyrophoric; Self-reactive; Organic peroxide; Self-heating; Emits flammable gases when in contact with water		Acute toxicity (harmful); Skin sensitizer; Irritant (skin and eye); Narcotic effect; Respiratory tract irritant; Hazardous to ozone layer (environment)
	Oxidizer		Respiratory sensitizer; Mutagen; Carcinogen; Reproductive toxicity; Target organ toxicity; Aspiration hazard
	Gas under pressure		Hazardous to aquatic environment
	Corrosive to metals		Acute toxicity (fatal or toxic)

Figure 2.34E GHS labels.

Information has also been supplied in the form of charts to provide supplemental information on situations when a tank may BLEVE and additional evacuation distances may be necessary (Figure 2.35A). Charts also contain some tactical information such as propane behavior characteristics and the amount of water needed to cool a tank. With the advent of terrorism over the past 25 years, additional information has been added to the guidebook to assist responders in determining if a hazardous materials incident could be an act of terrorism and provides information on terrorist agents and devices. An Improvised Explosive Device (IED) Safe Standoff Distance Chart is provided in the book as well. Educational information including a glossary has been provided in the white pages

Other Hazardous Liquids Pipelines

Some liquid pipelines transport highly volatile liquids that rapidly change from liquid to gaseous when released from a pressurized pipeline. Examples of these types of liquids include carbon dioxide, anhydrous ammonia, propane, and others.

Pipeline Markers

Since pipelines are usually buried underground, pipeline markers are used to indicate their presence in an area along the pipeline route. Of the three types of pipelines typically buried underground — distribution, gathering, and transmission — only transmission pipelines are marked with the following above-ground markers used to indicate their route.

Markers warn that a transmission pipeline is located in the area, identify the product transported in the line, and provide the name and telephone number of the pipeline operator to call. Markers and warning signs are located at frequent intervals along natural gas and liquid transmission pipeline rights-of-way, and are located at prominent points such as where pipelines intersect streets, highways, railways, or waterways.

Pipeline markers only indicate the presence of a pipeline—they do not indicate the exact location of the pipeline. Pipeline locations within a right-of-way may vary along its length and there may be multiple pipelines located in the same right-of-way.

NOTE:

- Markers for pipelines transporting materials containing dangerous levels of hydrogen sulfide (H_2S) may have markers that say: "Sour" or "Poison."
- Natural gas distribution pipelines are not marked with above-ground signs.
- Gathering/production pipelines are often not marked with above-ground signs.

Figure 2.34F Pipeline section.

near the end of the ERG to help responders understand how the book is organized.

The ERG is a valuable tool to be used to identify hazardous materials and determine actions to be taken by first responders in the first 30 minutes of a dangerous goods/hazardous materials incident. While the DOT stresses, the ERG is just for transportation incidents only. There is information in the ERG that can be useful for any hazardous materials incident, particularly if the ERG is the only information resource available in the initial phases of an incident. Just as first responders are limited in

Figure 2.35 CHEMTREC.

WARNING: The data given are approximate and should only be used with extreme caution. These times can vary from situation to situation. LPG tanks have been known to BLEVE within minutes. Therefore, never risk life based on these times.

BLEVE
(USE WITH CAUTION)

Capacity		Diameter		Length		Propane Mass		Minimum time to failure for severe torch	Approximate time to empty for engulfing fire	Fireball radius		Emergency response distance		Minimum evacuation distance		Preferred evacuation distance		Cooling water flow rate	
Litres	(Gallons)	Meters	(Feet)	Meters	(Feet)	Kilograms	(Pounds)	Minutes	Minutes	Meters	(Feet)	Meters	(Feet)	Meters	(Feet)	Meters	(Feet)	Litres/min	USgal/min
100	(26.4)	0.3	(1)	1.5	(4.9)	40	(88)	4	8	10	(33)	90	(295)	154	(505)	307	(1007)	94.6	25
400	(105)	0.61	(2)	1.5	(4.9)	160	(353)	4	12	16	(53)	90	(295)	244	(801)	488	(1601)	189.3	50
2000	(528)	0.96	(3.2)	3	(9.8)	800	(1764)	5	18	28	(92)	111	(364)	417	(1368)	834	(2736)	424	112
4000	(1057)	1	(3.3)	4.9	(16.1)	1600	(3527)	5	20	35	(115)	140	(459)	525	(1722)	1050	(3445)	598	158
8000	(2113)	1.25	(4.1)	6.5	(21.3)	3200	(7055)	6	22	44	(144)	176	(577)	661	(2169)	1323	(4341)	848	224
22000	(5812)	2.1	(6.9)	6.7	(22)	8800	(19400)	7	28	62	(203)	247	(810)	926	(3038)	1852	(6076)	1404	371
42000	(11095)	2.1	(6.9)	11.8	(38.7)	16800	(37037)	7	32	77	(253)	306	(1004)	1149	(3770)	2200	(7218)	1938	512
82000	(21662)	2.75	(9)	13.7	(45)	32800	(72310)	8	40	96	(315)	383	(1257)	1435	(4708)	2200	(7218)	2710	716
140000	(36984)	3.3	(10.8)	17.2	(56.4)	56000	(123457)	9	45	114	(374)	457	(1499)	1715	(5627)	2200	(7218)	3539	935

Figure 2.35A Information has also been supplied in the form of charts to provide supplemental information on situations when a tank may BLEVE and additional evacuation distances may be necessary.

what they can do at the scene of a hazmat emergency by equipment and training, the ERG is limited in the amount of information it provides. ERG was not intended to be used during the mitigation phase of the incident and therefore should not be used as one of the reference materials selected to determine mitigation methods. Free copies of the ERG are provided by the DOT to all fire, police, EMS and other emergency response organizations through a selected agency in each state.

PSMSA partnered with the National Library of Medicine (NLM) to provide a free smartphone version of the ERG 2020 for emergency responders. Copies are also available for a fee from private companies on line. The next edition of the ERG will be published in 2024.

UN/DOT Classification System for Hazardous Materials

Explosives—Class 1. The symbol is an exploding sphere on an orange background, and there are six subclasses:

1.1 Explosive with mass explosion hazard (detonation)
1.2 Explosive with projection hazard (detonation or deflagration)
1.3 Explosive with fire, minor blast or minor projection hazard
1.4 Explosive device with minor explosion hazard
1.5 Very insensitive explosives
1.6 Extremely insensitive explosives

Today placards, other than radioactive, do not contain the words designating the hazard class; they are referred to as "wordless placards."

Compressed Gas—Class 2. There are three subclasses: flammable gas, poison gas, or nonflammable nonpoisonous gas. The flammable gas placard is red with a flame symbol at the top. The nonflammable gas placard is green with a compressed gas cylinder symbol at the top. The poison gas placard is white with a skull and crossbones symbol at the top.

2.1 Flammable gas
2.2 Nonflammable gas
2.3 Poison gas

Flammable Liquid—Class 3. There are no subclasses in Class 3. Placards are red with a symbol of a flame at the top.
Flammable Solid—Class 4. Flammable solid, spontaneously combustible and dangerous when wet materials are included in this hazard class. The flammable solid placard has red and white vertical "candy" stripes with a flame at the top. The spontaneously combustible placard is white on the top half, and red on the bottom half, with a flame symbol at the top. The dangerous when wet placard is blue with a flame symbol at the top.

4.1 Flammable solid
4.2 Spontaneously combustible
4.3 Dangerous when wet

Oxidizer—Class 5. This class includes two subclasses: oxidizer and organic peroxide. These placards are yellow with the symbol of a flaming sphere at the top.

5.1 Oxidizer
5.2 Organic peroxide

Poison—Class 6. This class includes two subclasses: poison and infectious substance. The poison placard is white with a skull and crossbones at the top. There are also variations of this placard such as stowing away from foodstuffs, with a stalk of wheat with an "X" through it and a half placard with "Marine Pollutant" on it.

6.1 Poison
6.2 Infectious substance

Radioactive—Class 7. Placards in this class contain materials that carry the Radioactive III label. The placard is white on the top half and yellow on the bottom half. The radioactive propeller symbol is located at the top of the placard. The word "Radioactive" **MUST** appear on the placard. There are no subclasses.

Corrosive—Class 8. This class contains both acids and bases. The placard is white on the top half and black on the bottom half. The hand and metal corrosion symbol is at the top. There are no subclasses.

Miscellaneous Hazardous Materials—Class 9. This class contains materials not covered by other classes. The placard has black and white vertical stripes on the top half with a white bottom half. There is no wording on the placard. There are no subclasses.

Dangerous Placard. This placard indicates a mixed load of other hazard classes, excluding Table 1 materials. The placard has a red background with a white center portion with the word "Dangerous."

Placard Requirements
Table 1

The following hazard classes of materials, also known as Table 1 materials, **must** be placarded on the container, trailer or truck in highway transportation regardless of the quantity of material being shipped:

1. Explosives 1.1
2. Explosives 1.2
3. Explosives 1.3
4. Poison Gas 2.3
5. Dangerous When Wet

6. Poison 6.1 (Inhalation Hazard Only)
7. Radioactive (III Label Only)

Table 2

All other hazard classes (Table 2 materials) are placarded when the weight of any of the hazard classes reaches 1,001 lb or more, or when the aggregate weight of two or more reaches 1,001 lb or more. This applies only to highway transportation.

Stenciled Commodities

Some hazardous materials because of their danger have the name of the material stenciled on the side of the tank truck. Some Class 2.3 poison gases and 6.1 liquids, which are inhalation hazards, require that the words "INHALATION HAZARD" appear along with the stenciled name.

DOT Chart 16

DOT Chart 16 has colored pictures of all of the different placards and labels used in the transportation of hazardous materials. The chart has a partial listing of some of the transportation regulations from CFR 49 that pertain to placard and labeling (Figure 2.36). Chart 16 also contains two tables that identify which hazardous materials must be placarded under what conditions. Table 1 materials must be placarded regardless of the quantity of material being transported. Table 2 materials must be placarded when there are 1,001 lb or more of hazardous materials on board. A black and white copy of Chart 16 is included in the appendix and can be very helpful during emergency response and in conducting commodity flow studies of transportation containers and routes. These charts are issued and updated by DOT periodically. Chart 16 is the latest of 16 versions so far. Periodically check the DOT website for updated versions.

DOT Placard and Label System

First responders should become familiar with the DOT hazard classes for hazardous materials and the placards and labels used to identify those hazards. This information will assist the first responders when making an initial recognition and using the ERG. There are nine hazard classes identified by DOT. Each hazard class has a particular color associated with its placard and label. Placards and labels are diamond shaped. The placard is the largest and is used on the outside of transportation vehicles and certain large containers. Labels are much smaller and found on the individual packages and small containers of hazardous materials. Also located on each placard and label is the hazard class number. This number is located on the bottom corner of the diamond. A symbol is found in the top corner of the diamond, and on most of the placards and labels, a hazard class name is located in the center. DOT regulations permit the

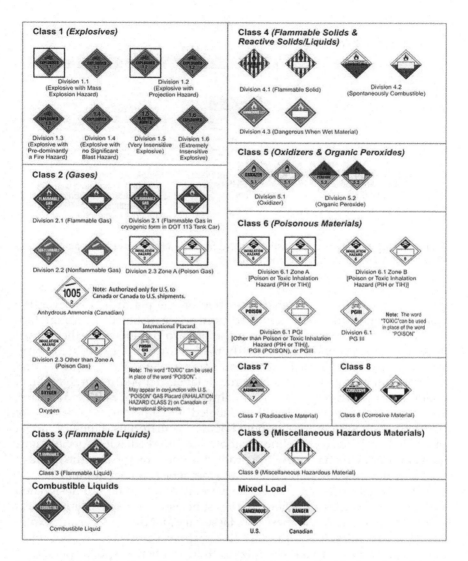

Figure 2.36A Placards and labels are diamond shaped with the placard being the largest and used on the outside of transportation vehicles and certain large containers.

use of wordless placards, which are used throughout the rest of the world. Wordless placards do not have the hazard class name in the center of the placard. Responders should become familiar with the colors of the placards and labels and the hazard class associated with the color. Many times the nature of hazardous materials incidents does not allow personnel to

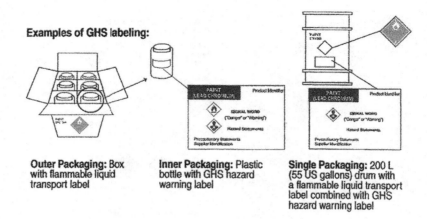

Examples of GHS labeling:

Outer Packaging: Box with flammable liquid transport label

Inner Packaging: Plastic bottle with GHS hazard warning label

Single Packaging: 200 L (55 US gallons) drum with a flammable liquid transport label combined with GHS hazard warning label

Figure 2.36B Labels are much smaller and found on the individual packages and small containers of hazardous materials.

get close enough to read what is written on the placards. However, the color of the placard can be identified from a safe distance or through binoculars. By associating the color of the placard or label with the hazard class, the first responder can identify the hazard even if nothing is written on the placard. Hazardous materials transported by rail, highway and waterway may be found with placards or labels depending on the type of container or packaging. Usually, air shipments present a limited hazard, are in small packages and will be labeled. Pipelines are also a type of transportation system for shipping hazardous materials. Pipelines aren't placarded or labeled. Pipeline locations are marked with special signs aboveground, which identify the product in the pipeline and the name of the pipeline company, and provide an emergency contact number.

Orange placards indicate explosive materials, which are designated as DOT hazard Class 1. There are six subclasses of explosive materials identified by DOT. While these subclasses are identified by certain explosive characteristics, it is important that response personnel do not relate any of the classes with a lesser hazard than another. Response personnel may not know if the circumstances are present for an explosive material to behave in a certain way. Therefore, all explosive classes should be treated as the worst case, which is detonation, and treat them accordingly, until explosive experts arrive on scene to deal with the materials. Red placards indicate that the hazardous materials are flammable.

Red placards may be found in hazard Class 2 compressed gases with a flammability hazard and Class 3 flammable liquids. Flammable gases include propane, hydrogen and butane. Flammable liquids include gasoline, acetone, alcohols and ketones. When response personnel recognize that a flammable hazard exists, they should take every precaution to prevent ignition of the material. Ignition sources can include open flames, smoking, welding and other hot operations, heat from friction, radiant heat, static electrical charge, electrical sources, mechanical sparks and spontaneous ignition. Fire apparatus can be a source of ignition and should be positioned properly to avoid being an ignition source. Compressed gases are a hazard class because of the pressure in the containers, which presents a hazard in addition to the physical and chemical characteristics of the gases. A green placard also indicates a compressed gas, which is considered by DOT to be nonflammable. This can be dangerously deceiving, because anhydrous ammonia is placarded as a nonflammable compressed gas, when, in fact, it will burn under certain conditions, usually inside a building or in a confined space. This is because the DOT definition of a flammable gas does not fit the flammable range of anhydrous ammonia. Other nonflammable gases include carbon dioxide, nitrogen and argon. There is also a white compressed gas placard for poisons. Poison gases include chlorine and phosgene.

Flammable solid materials are hazard Class 4 and have three different placards based upon general hazards. The first is flammable solids with a red and white striped placard. An example of a flammable solid would be highway flares. Next are materials which are spontaneously combustible with a white over red placard. Even though the hazard class is flammable solids, spontaneously combustible liquids are also included because they don't fit anywhere else in the hazard class system. An example of a spontaneously combustible material would be phosphorus. The third flammable solid placard is dangerous when wet which is all blue. Dangerous when wet or water reactive materials as they are also known include sodium, potassium and other metals. Class 5 materials are oxidizers and have yellow placards. Oxidizers include swimming pool chemicals and oxygen.

Class 6 materials are liquid and solid poisons with white placards. These include liquids, which are very volatile, but not gases, and infectious substances. Military and terrorist nerve and mustard agents and biological materials would be included in this category. Poison gases are placed in Class 2. Radioactive placards are Class 7 and are yellow over white. There are three radioactive labels placed on packages according to the amount of radiation emitted by the materials. Class 8 materials are corrosive with a white over black placard. Corrosive materials can be acids or bases and include sulfuric acid and potassium hydroxide. Class 9 materials are miscellaneous hazardous materials with black and white stripes over white. Miscellaneous materials include sulfur and hazardous waste. DOT regulations, including the hazard classes, placards and labels, undergo periodic review and revision. Response personnel need to watch for changes in transportation regulations that may affect them during an emergency response or when identifying hazardous materials.

Fixed facilities may also use the DOT placard and labeling system for storage and use of hazardous materials at their facility. OSHA requires that all placards and labels in place during transportation remain in place during storage and use. The markings can only be legally removed once a container has been purged of the hazardous materials and properly disposed of. The National Fire Protection Association (NFPA) also has a marking system **used for** fixed facilities storing or using hazardous materials. Unlike the DOT and OSHA regulations, NFPA 704 is an optional standard and does not carry any legal weight unless adopted by a local jurisdiction.

Placards of varying sizes are used in the system to identify major hazards of chemicals in the facility. They do not identify specific chemicals, but rather generic hazard classes. The hazard classes by color are not all the same as the DOT hazard class colors. Each NFPA 704 placard includes four colored sections, which indicates a particular hazard. These hazards include a blue section for health, a red section for flammability, a yellow section for reactive and a white section for miscellaneous information such as water reactivity, corrositivity and radioactivity. Within the colored sections of the placard, numbers 0, 1, 2, 3 **or** 4 are **placed** to indicate the level of severity of the hazard that the material in the hazard area presents. A 0 in one of the colored sections indicates no hazard, and a 4 indicates the most severe hazard. There are degrees of hazard associated with each of the three colors on the 704 placard, and responders should become familiar with the designations.

Private Sector Resources and Regulations

Placard Hazard/Chemistry Chart

Most hazardous materials have more than one hazard. DOT placards are only required for the most severe hazard of a material (Figure 2.37).

> **Author's Note:** *About 20 years ago, because of the short comings in terms of identification of hazards, I employed chemistry to develop a Placard Hazard Chart. By including information from chemistry with the DOT's most significant hazard of a given material as represented by the placard, most all materials hazards can be identified until more specific identification information becomes available. Perhaps the advent of the Study of the Science of Hazardous Materials may lead to additional information being available for identification of an unknown material, or at the very least, its hazard or multiple hazards, and result in a safer and quicker resolution of a hazardous materials incident.*

Burke Placard Hazard Chart

X-Primary Hazard
* Secondary Hazard

Placard Color	Explosive	Fire Hazard	Water Reactive	Air Reactive	Toxic	Corrosive	Compressed Gas	Thermal	Polymerize	Spontan Comb	Asphixiant	Oxidizer	Heat/Shock Sen	Radioactive
Orange	X	*			*	*						*	*	
Red		X			*	*	*	*Cold	*		*			
Green	*				*	*	X	*Cold				*	*	
Red & White Stripes	*	X	*	*	*									
White Over Red	*				*	*				X		*		
Blue		*	X	*	*	*								
Yellow / Red Over Yellow	*			*	*	*	*	*Cold				X		
White		*	*		X	*	*	*Cold	*			*		
Yellow Over White					*					*		*		X
White Over Black		*	*		*	X					*	*		
**Black & White Stripes Over White	*	*	*	*	*	*	*	*Hot	*	*	*	*	*	*
Red White Red	*	*	*	*	*	*	*	*Cold	*	*	*	*	*	*

© 2020 RAB   ** Class 9 includes hazardous waste, which can be composed of materials from many of the other hazard classes.

Figure 2.37 Most hazardous materials have more than one hazard. DOT placards are only required for the most severe hazard of a material.

Fixed Facility Marking

In addition to transportation, hazardous materials are stored and used in every community in the United States and many other countries. Responders should be able to identify potential locations of hazardous materials in the community by the type of occupancy and location where they may be stored or used. For example, in many small communities across the country, farm chemical supply stores may have pesticides, anhydrous ammonia, propane, ammonium nitrate, motor vehicle fuels and other chemicals. So, if a farm chemical supply store is located in your community, expect to find those types of chemicals there. Hospitals may contain liquid oxygen, anesthetic gases, cleaning supplies and disinfectants, to name a few. Water and sewage treatment plants may contain chlorine and other treatment chemicals.

High schools and colleges can contain many different kinds of chemicals in science class rooms, flammable liquids and gases in shop areas and cleaning supplies. The list can go on and on. Responders can conduct preplanning inspections of fixed facilities to determine what chemicals are stored and used there. First responders should also be familiar with occupancies and locations that could become terrorist targets using hazardous materials. These might include government buildings, places where large numbers of people assemble, public transportation and infrastructure, telecommunications facilities and public utilities, and historical locations. Fixed facilities may also use the DOT placard and label system for storage and use of hazardous materials at their facility. OSHA requires that all placards and labels in place during transportation remain in place during storage and use. The markings can only be legally removed once the container has been purged of the hazardous materials and properly disposed.

NFPA 704 Fixed Facility Marking System NFPA 704 or Hazard Communication should be used to identify areas within buildings where dangerous chemical or biological materials may be used and stored along with emergency contact information for use by response personnel during an emergency (Figure 2.38). Some chemicals found in laboratories are harmless, while others can be flammable, oxidizers, corrosive, toxic or explosives. Firefighters may be confused by chemical names and not know by the name if a material is dangerous or not. Warning markings on smaller containers are not always conspicuous. While positive identification is most desired, determining a hazard class or chemical family can identify the hazard generally associated with the chemical. The chart on "Hints for Hazardous Materials" can be helpful in determining relative danger of chemicals as an initial starting point. The *North American Emergency Response Guidebook* (NAERG) can also be useful to first responders. Listed below are some common laboratory chemicals which become increasingly dangerous as they age.

Figure 2.38 NFPA 704 or Hazard Communication should be used to identify areas within buildings where dangerous chemical or biological materials may be used and stored along with emergency contact information for use by response personnel during an emergency.

NFPA 704 Hazardous Materials Marking System Most everyone involved with hazardous materials response is familiar with the DOT placard and label system for transportation of hazardous materials. OSHA has also adopted the DOT system and requires hazardous materials placarded in transportation to continue to be placarded during storage and use. Placards and labels must remain in place until the material is used up and the container is purged of the residue. There are also other fixed storage and use marking systems, including the NFPA 704 system, OSHA Hazard Communication Systems and miscellaneous internal marking systems for occupancies, which store and use hazardous materials.

Work on the development of NFPA 704 started in 1952, which resulted in the establishment of the standard in 1957. This system was one of the first efforts to identify locations where hazardous materials were stored and used, but does not apply to transportation. DOT did not develop their system for transportation marking for another 20 years. Because 704 is a standard, it is not required or mandated unless a local jurisdiction adopts the standard and makes it law within the community.

Usage of 704 varies from community to community. Some facilities may place 704 diamonds on buildings and locations where hazardous materials are stored or used voluntarily (Figure 2.38A). So it is quite possible that 704 diamonds could be found even in communities where it has not been adopted or required by law. NFPA 704 is a generic hazardous materials marking system designed to alert emergency responders to the

Figure 2.38A Usage of 704 varies from community to community. Some facilities may place 704 diamonds on buildings and locations where hazardous materials are stored or used voluntarily.

presence of hazardous materials at a particular occupancy, to assist them in evaluating the hazards present and to help them in planning effective fire and emergency control operations. NFPA 704 does not provide specific information about individual chemicals. The information won't provide chemical names. It is a basic identification system to help first responding emergency personnel to decide whether to evacuate the area or to commence control procedures. It also assists responders with selection of firefighting tactics, appropriate personal protective equipment and emergency procedures.

NFPA 704 is intended to provide information to emergency responders about the general hazards of materials that may be inside an occupancy. The hazards on a diamond do not provide information about every hazardous material in a facility (Figure 2.39). NFPA 704 diamonds list the most severe hazards of the most hazardous chemicals present. They are really only a "stop" sign to warn responders and to allow them to be aware of hazardous materials present. Information provided will not indicate the routes of entry for toxic materials, degrees of radioactivity, corrosivity or other specific chemical information. More information will be needed before mitigation efforts are undertaken.

NFPA 704 "diamonds" are placed on the outside of buildings near the entrances or addresses on the buildings. They may also be placed on the inside of the building where the hazardous materials are actually stored and used. Exact locations are left up to the Authority Having Jurisdiction (AHJ), but minimum requirements in the standard require the following locations:

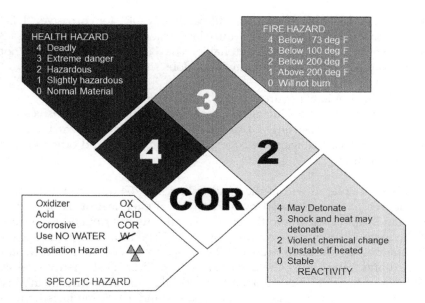

Figure 2.39 NFPA 704 is intended to provide information to emergency responders about the general hazards of materials that may be inside an occupancy. The hazards on a diamond do not provide information about every hazardous material in a facility.

- Two exterior walls or enclosures containing a means of access to a building or facility
- Each access to a room or area
- Each principal means of access to an exterior storage area

Diamonds used on the outside of buildings are generally a minimum size of 15 in. × 15 in. Those used inside of buildings are a minimum of 10 in. by 10 in. Smaller sized diamonds may be found on individual containers. NFPA 704 placards are divided into four colored quadrants, or four "mini" diamonds within the larger diamond. The upper quadrant is red in color at the "12 o'clock" position followed by blue at "9 o'clock" to the left and yellow at "3 o'clock" to the right, and the bottom is white at the "6 o'clock" position. These colors designate a specific hazard. Red indicates flammability, blue health, yellow instability and white special information.

Special information located in the white quadrant listed in the 704 standard includes materials that react violently or explosively with water, represented by a "W" with a slash through it and an "OX" for materials with oxidizing properties. The standard lists specific criteria for these symbols to be included on the diamond in the appendix. Other users of the 704 system have added other symbols, which are not a part of the 704 standard, and include "RAD" for radioactive, "COR" for corrosive, "UD" for unclassified detonable, "4D" for Class 4 detonable, "3D" for Class 3 detonable and "3N" for non-detonable. The NFPA 704 standard is not intended to speak to the following hazardous situations:

- Occupational exposures
- Explosive and blasting agents, including commercial explosives
- Chemicals whose only hazard is as a chronic health hazard
- Teratogens, mutagens, oncogens, etiologic agents and other similar hazards

Numbers are placed within the quadrants of the 704 diamond indicating the degree of hazard posed by an individual chemical or as quoted in the standard, "according to the ease, rate, and quantity of energy release of the material in pure or commercial form." The numbers used to identify the range of hazards are from 0 to 4, with 0 indicating no particular hazard and 4 indicating the most severe hazard or most energy release. Degrees of hazard associated with health, flammability and instability are determined by criteria outlined in the standard. Some jurisdictions may place a letter "G" in the quadrant with a number to indicate the presence of a compressed gas.

Health hazards are ranked according to the level of toxicity and effects of exposure to response personnel. They are based upon short-term, acute exposure during handling under conditions of spill, fire or similar emergencies. Short-term exposure ranges from minutes to hours. Acute exposures typically are sudden and severe, and characterized by rapid absorption of the chemical that is quickly circulated through the body and damages one or more vital organs. Acute effects include severe burns, respiratory failure, coma, death or irreversible damage to a vital organ.

Author's Note: The following guidelines are my effort to explain the material in NFPA 704 in terms of protection of response personnel. They are not intended to be recommendations. Materials at the scene of a hazardous materials incident should be thoroughly researched, and decisions made by those on scene as to the appropriate level of protection responders should wear based upon the hazards present. Specific details of criteria used to determine hazard numbering are located in the NFPA 704 standard.

- **4**—Materials that can be lethal if response personnel do not wear proper chemical protective equipment. If gases or skin absorbent gases, liquids or solids, Level A chemical protection would be necessary. Firefighter turnouts would not provide appropriate protection. Some examples of Health Hazard 4 chemicals include chlorine, phosgene, hydrocyanic acid (hydrogen cyanide), hydrogen sulfide, phenol, phosphine, pentaborane and acrylonitrile (vinyl cyanide).

- **3**—Materials that can cause serious or permanent injury. Level A chemical protection or Level B chemical protection would be appropriate depending on the physical state of the hazardous materials and availability of monitoring instruments. Firefighter turnouts would not provide appropriate protection. Some examples of Health Hazard 3 include anhydrous ammonia, acetaldehyde, acrylic acid, carbon monoxide, formic acid, pyridine, nitric acid and *para*-xylene (*p*-xylene).
- **2**—Materials that can cause temporary incapacitation or residual injury. Level B or Level C chemical protection would be appropriate for these chemicals. Firefighter turnouts would likely not provide appropriate levels of protection if no fire or threat of fire exists. Health Hazard 2 chemicals include *meta*- and *ortho*-xylene (*m*-xylene and *o*-xylene), toluene, styrene, ethyl formate, benzene, 1,1,1-trichloroethane and vinyl chloride.
- **1**—Materials that can cause significant irritation. Level C chemical protection would likely be appropriate protection. Firefighter turnouts may provide some protection, particularly respiratory. Many irritants are actually solid materials and may contaminate personnel. Remember that firefighter turnouts are not classified as chemical protective clothing; they should only be worn if there is danger of fire and a significant reason for fighting the fire, not just because we can. Health Hazard 1 chemicals include, acetone and butane.
- **0**—Materials that would offer no hazard beyond that of ordinary combustible materials. Firefighter turnouts would provide appropriate protection for personnel.

Flammability hazards are based upon a material's susceptibility to burning. Conditions present need to be considered as well as the combustibility characteristics of the fuel. Firefighter turnouts are generally the appropriate protective clothing for flammability hazards. There may, however, be conditions where firefighters even in full turnouts cannot be adequately protected because of the volume of fire or flame impingement on containers. Each situation needs to be thoroughly evaluated based upon the flammability hazards present. Some flammable chemicals may also be toxic, and toxicity should be taken into account before emergency personnel are sent into a scene where toxic materials may be present or on fire.

Instability hazards (reactivity) address the degree of inherent vulnerability of materials to release energy. They apply to materials capable of rapidly releasing energy by themselves, through self-reaction or polymerization. They do not deal with water reactive materials. Organic peroxides need to be evaluated with NFPA 432 Code for the Storage of Organic Peroxide Formulations. Instability does not take into account the unintentional combination of materials, which may occur during fire or other conditions. During storage, unintentional mixing should be considered in order to establish appropriate separation or isolation. The degree of instability hazard is meant to indicate to emergency personnel if an area should be evacuated, if firefighting should be conducted from a location of cover, if caution is required when approaching for extinguishment based upon the extinguishing agent or if a fire can be fought using normal procedures. Ranking of instability hazards is based upon a material's ease, rate and quantity of energy release.

- **4**—Materials sensitive to shock and heat at normal temperatures and pressures. They can undergo detonation or explosive decomposition at normal temperatures and pressures. Examples of Instability Hazard 4 include ammonium perchlorate, 3-bromopropyne, chlorodinitrobenzenes, fluorine, nitromethane, peracetic acid and picric acid.
- **3**—Materials sensitive to shock and heat at elevated temperatures and pressures. They are subject to detonation or explosive decomposition or explosive reaction, but need a strong initiating source or must be heated under confinement. Examples of Instability Hazard 3 include tetrafluoroethylene, silane, *n*-propyl nitrate, perchloric acid, nitroethane, hydrogen peroxide >60%, hydrazine, ethylene oxide, chloropicrin and ammonium nitrate.
- **2**—Chemicals that will undergo violent chemical change at elevated temperatures and pressures. Examples of Instability Hazard 2 chemicals include trichlorosilane, titanium tetrachloride, sulfuric acid, styrene monomer, sodium hydride, potassium, phosphorus trichloride, phosphorus, methyl isocyanate, hydrogen cyanide, ethylene, epichlorohydrin and aluminum chloride.
- **1**—Materials that in themselves are normally stable, but that can become unstable at elevated temperatures and pressures. Examples of Instability Hazard 1 materials include acetic anhydride, allyl alcohol, aluminum powder, ethyl ether, hydrogen chloride, magnesium, nitrobenzene, phosgene and potassium hydroxide.
- **0**—Materials that themselves are normally stable, even under fire conditions.

Special hazards listed in the text of NFPA 704 are limited to water reactive and materials with oxidizing properties, which would require special firefighting techniques. Special hazards are indicated by a symbol located in the quadrant of the diamond at the 6 o'clock position in the white colored section. Materials that react violently or explosively with water are identified by the letter "W" with a line through the center. Materials that possess oxidizing properties are identified by the letters "OX". Examples of chemicals with an **"OX" hazard** include sodium peroxide, potassium peroxide, peracetic acid, oxygen, nitric acid, hydrogen peroxide, chromic acid, bromine and ammonium nitrate. **Water reactive** materials include calcium carbide, calcium hypochlorite, fluorine, lithium aluminum hydride, potassium and sodium.

NFPA 704 is just one of many tools available for emergency responders when evaluating an incident scene for the presence of hazardous materials. More information needs to be gathered before mitigation tactics are decided upon. Generally, firefighter turnouts do not provide much protection against hazardous materials, although self-contained breathing apparatus (SCBA) provides a high level of respiratory protection. A thorough evaluation of the hazards present along with a risk benefit analysis must take place before personal protective equipment and tactics are selected.

Military Placard System
The military also has a placard system for explosive materials, chemical agents and biological materials (Figure 2.40). The system uses orange placards of various shapes to indicate levels of fire and explosion

Military Explosives Placards

| Class 1 Division 1 Mass Detonation Hazard | Class 1 Division 2 Explosion and Fragmentation Hazard | Class 1 Division 3 Mass Fire Hazard | Class 1 Division 4 Moderate Fire Hazard |

Military Chemical & Firefighting Symbols

| Highly Toxic Chemical Agents | Harassing Agents | White Phosphorus Munitions | Wear Protective Mask or Breathing Apparatus | Apply No Water |

Figure 2.40 The military also has a placard system for explosive materials, chemical agents and biological materials.

hazards of materials. There are also a series of circular placards with chemical suits and protective clothing, which identify chemical and biological hazards. Red symbols indicate highly toxic chemical agents, yellow indicate harassing agents and white indicate white phosphorus. Also included in the system is a circular water reactive placard. If your jurisdiction is located near a military facility, or has mutual aid with one, you should be familiar with the military marking system.

Additional Resources

Shipping Papers

Shipping papers can be found with four of the five DOT transportation systems. With railroad shipments, the shipping papers are called the Way Bill or Consist and located in the train engine, usually under the control of the conductor. Most trains no longer have a caboose for the conductor to ride in. Highway shipments have shipping papers called the Bill of Lading or Freight Bill and are located in the cab of the truck, usually in a slot in the driver side door. Water transportation vessels have shipping papers called the Dangerous Cargo Manifest, which is located in the wheelhouse. If a barge is involved, the shipping papers are also found in a small container on the barge. Barges may also have a red pennant flying, indicating that it is carrying hazardous materials. Even though the pennant is red, the color doesn't identify a particular hazard class. Aircraft transporting hazardous materials have shipping papers called the air bill and are located in the cockpit of the aircraft with the pilot. A simulation of a shipping paper can be found inside the front cover of the ERG, with an example of the type of information it contains.

Typical shipping papers provide an emergency contact number for the manufacturer of the hazardous material, number and type of packages, chemical name, hazard class, quantity of product and UN identification number. Quantities will be listed using the metric system of weight, so response personnel should have a working knowledge of the metric system. The shipping papers may also identify a quantity of material, which must be reported to the NRC if released, and will be marked with an "RQ", indicating a reportable quantity. Materials with an "RQ" are on the EPA list of materials that must be reported if released into the environment. The emergency contact phone number at the top of the first page of a shipping paper should be called for response information rather than CHEMTREC. If the shipping papers are not available, then CHEMTREC should be called.

Material Safety Data Sheets/Standard Operating Guidelines

MSDSs/(SOGs) are found at fixed facilities storing or using hazardous materials (Figure 2.41). Certain facilities with extremely hazardous

Figure 2.41 Captain Francisco Martinez of Lincoln Fire & Rescue in Nebraska looks over MSDS to gain information about a chemical before determining incident objectives. (Courtesy of Lincoln Fire and Rescue.)

materials identified by the EPA and LEPC are required to submit MSDSs to the local fire department, and information about the facility is available from the LEPC. The LEPC is created as a result of the Emergency Planning and Community Right-to-Know Act of 1986, also known as EPCRA. This act allows the local fire department to access any facility in the community that stores or uses hazardous materials for the purpose of preplanning a hazardous materials release. MSDSs contain chemical names, synonyms, hazards, protective equipment, spill cleanup information and decontamination information.

CHEMTREC

Do You Have an Emergency Involving Chemicals?
CHEMTREC **the Chemical Transportation Emergency Center**, began its operations on September 5, 1971. **CHEMTREC** is a 24-h emergency center that was started as a service of the Chemical Manufacturers Association (CMA), now known as the American Chemistry Council. The first center was located in downtown Washington, DC. It is now located across the Potomac River in Arlington, VA. In addition to its chemical emergency call center, CHEMTREC provides training programs, MEDTREC (for chemical medical information), **Material Safety Data Sheets (MSDS)** and CHEMNET, an industry chemical emergency mutual aid system.

Emergency Call Center

During an interview with Director of the Operations Center Joe Milazzo, it was interesting to find that the call volume to the Emergency Call Center is down. They currently handle 99,000 cases and 55,000 actual emergency calls per year, an average of 350 calls per day. He believes the reason is the easy access to online information through smartphones and internet. It is important to note that CHEMTREC should still be your first call, before searching electronic resources. There are a number of advantages to calling them first. One call to 800-424-9300 can put responders in touch with the company that manufactured a chemical, the shipper, provide MSDS, activate and send a chemical company or industry mutual aid hazmat response team to the scene. CHEMTREC can also provide on-scene responders with information on over 1 million chemicals. Procedures are in place for calls originating from non-English-speaking callers. Specialized assistance from chemists can be obtained from its extensive database of chemical industry contacts. Emergency collect calls are accepted, and all calls are recorded.

CHEMTREC provides emergency chemical information to emergency responders nationwide. Its primary mission "is to provide technical information about products involved, guidance on how to protect themselves and the public, and what initial action is required to mitigate the incident." CHEMTREC strives to provide this and other information quickly and accurately 24 h a day. Emergency response organizations do not have to be registered with CHEMTREC to use its services. Information and assistance are provided free of charge as a public service to emergency response organizations. Backup power is provided for the center, and protocols are in place for relocating the center in an emergency.

CHEMTREC History

In the beginning, CHEMTREC communicators, as they were called, were merely conveyers of information. Many were former military personnel who were recruited because of their ability to keep cool during crises. Initially, chemical data were kept in a sophisticated card file system, and the communicator would read the information on the card with little if any additional input (Figure 2.42). They were not required to have any expertise in the areas of hazardous materials or emergency response.

- **1918**

    Responding to a series of railway accidents involving shipments of corrosive liquids vital to war efforts, the Manufacturing Chemists' Association (MCA) forms a committee devoted to improvement of liquid chemical shipping containers.

Figure 2.42 Initially, chemical data were kept in a sophisticated card file system, and the communicator would read the information on the card with little if any additional input.

- **1932**

    MCA begins a comprehensive program of Safety Manuals and Safety Data Sheets. Each data sheet provides details of a product's physical and chemical properties, its hazards, instructions on safe handling, **first-aid, required labeling or identification** and methods of unloading and emptying various types of containers.
- **1969**

    The **U.S. Department of Transportation** met with MCA to determine the best approach for reporting and responding to emergency situations involving chemicals in transport.
- **1970**

    MCA established CHEMTREC **(CHEMical TRansportation Emergency Center)**, to provide chemical-specific information to emergency responders around the clock. Members authored over 1,600 chemical cards representing the chemicals most frequently involved in transportation.
- **1971**

    On September 5, CHEMTREC became fully operational.
- **1978**

    MCA became the **Chemical Manufacturer's Association. (CMA)**
- **2000**

    CMA became the American Chemistry Council (ACC).
- **2004**

    CHEMTREC launched its medical service coverage.

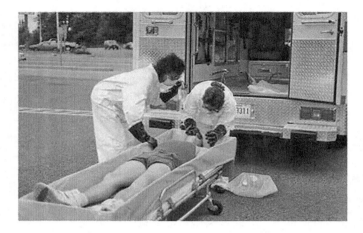

Figure 2.43 CHEMTREC provides information to emergency medical personnel concerning chemical exposure treatment.

Today's version of the communicator is the Emergency Services Specialist (ESS), highly trained and qualified in hazardous materials and emergency response. ESS personnel come from the emergency services with experience in emergency response, firefighting, emergency medical services, military emergencies and related fields. The old card file system has been replaced with a computerized database, and **Emergency Services Specialists** provide additional information and recommendations based on their knowledge and experience dealing with hazmat emergencies.

CHEMTREC can provide technical information directly to the scene of a hazmat incident, including faxed information and MSDS from its document library. CHEMTREC also provides information to emergency medical personnel concerning chemical exposure treatment (Figure 2.43). ESS personnel gather information about the incident, product and exposure, and then provide medical treatment information from the MSDS.

You can call CHEMTREC right after you arrive on scene, and they will be able to assist you. However, in order for CHEMTREC to provide information to the response scene in a timely manner, it is suggested that responders gather and have available information needed by CHEMTREC before making a call to them. The following is some of the information suggested:

- Caller's name and rank or title
- Caller's company or organization
- Caller's location
- At least one callback number, with the area code
- Dispatch center number, if available

- Fax number or email address
- Location of incident/weather conditions
- Time incident occurred
- Type or description of container/package
- Container numbers and/or markings
- Brief description of incident and actions taken
- Number and type of injuries/exposures
- Amount of product(s) involved and released
- Is there specific information needed as a priority?
- Are any industry representatives on the scene or have any been contacted?

   If shipping papers have been obtained, the following additional information is requested by CHEMTREC:

   a. UN/NA identification number (placard) or Standard Transportation Commodity Code (STCC) number of the products
   b. Chemical name, product(s) name or a trade name
   c. Carrier name
   d. Shipper and point of origin
   e. Consignee and destination

Not all of the incidents received by CHEMTREC are massive train derailments or tank truck accidents. As many as 50% of the incidents called into the emergency center involve 5 gallons or less of product. Below are examples of two typical and one unusual incident handled by the call center. (Sample CHEMTREC incidents are from the User's Guide for Emergency Responders.)

   Incident 1. CHEMTREC received a call from a railroad reporting a train derailment. The report stated that six cars containing sulfuric acid and two empty cars that had carried naphtha/xylene mixture and silicon tetrachloride were involved. An unknown amount of product was reported to be leaking from the railcars. Adverse weather conditions in the area were hampering response and evaluation of the incident. CHEMTREC faxed MSDS, and then discussed the situation with the local fire department and the nearest poison control center. CHEMTREC was also able to conference with the shippers, the fire department and railroad representatives.

   Incident 2. The trucking company dispatcher called in to report that an unknown product was leaking out from the trailer at a truck stop in Arizona. The local fire department was notified and contacted CHEMTREC for assistance identifying the products in the trailer. As there were multiple shippers' products on board, CHEMTREC asked that the bill of lading be faxed to the Emergency Call Center. The shipping papers received indicated that two shippers were involved, shipping hydrofluoric acid solutions and ethylene glycol. The ESS contacted each shipper, requesting contact with the reporting carrier's dispatcher. In turn, shippers called back to advice that their respective emergency coordinators had made contact and that product information had been supplied to responders.

Incident 3. While the author was visiting CHEMTREC to obtain information for this column, a call was received from a state National Guard and logged by an ESS. An auctioneer found a container marked as chemical weapon gas. The container was also marked "Manufactured by Lake Erie Chemical." A bomb squad had already entered the building and was able to view the container. The container was described as a plastic cylindrical object approximately 15 in. in length. There were two silver vials inside with a percussion cap. There was a brown opaque liquid in the vials. Cotton padding was in place between the two vials. The container was marked as "Lot 2250."

The ESS researched internet websites to obtain the following information, which was sent on to the National Guard. During the 1930s, the Lake Erie Chemical Company, based in Cleveland, developed a system to discharge tear gas into bank lobbies during holdups. This was done in conjunction with a company called Diebold, which was run by Elliot Ness, of "The Untouchables" fame. Lake Erie Chemical also manufactured flare pistols. One of the items in the tube was described as having a percussion cap on its base. CHEMTREC advised the National Guard of the strong possibility that the item was a dispensing unit that may contain tear gas. The item was collected and secured for disposal by the local bomb squad.

Assistance for Responders

Shipping Paper Emergency Contact Information The **U.S. Department of Transportation (DOT)** requires that a shipper of hazardous materials provide a 24-h emergency contact number at the top of all shipping papers. If a company does not have the capability of providing this type of contact, CHEMTREC can contract with the company and provide the emergency contact service. Therefore, some shipping papers will have the CHEMTREC 24-h emergency number at the top for emergency contact information. Others will have a number directly connected to the shipper or manufacturer. Response personnel should use the shipper or manufacturer number first, if provided, before contacting CHEMTREC.

MSDS System CHEMTREC has one of the largest databases of **Material Safety Data Sheets** in the world. Over 2.8 million documents are available and can be faxed to emergency response personnel on the scene of a hazmat incident. The MSDS system at CHEMTREC was recently upgraded to allow for better management and quicker retrieval of documents during an emergency. Information can be faxed directly to the incident scene.

Response personnel should identify the location of the nearest fax machine and provide the number to CHEMTREC in order to receive information.

Participation in Drills and Exercises

Through prior arrangement, CHEMTREC will participate in local hazmat drills and exercises. Using the Emergency Call Center during a drill or exercise can help emergency responders to better understand the resources and services available during an actual emergency. Participation by product shippers or manufacturers may also be arranged for specific chemicals.

MSDS for materials involved in a drill or exercise can be obtained prior to the event by contacting CHEMTREC's nonemergency number. Request forms for CHEMTREC participation are available by calling 800-262-8200 or downloading from the website. Completed forms should be faxed or emailed at least 48 **hours** prior to the drill or 10 days prior for regular mail. Mail should be sent to CHEMTREC Emergency Call Center, 1300 Wilson Blvd., Arlington, VA 22209-2380. Call 703-741-5525 at least 1 day in advance of the drill to confirm that the registration was received.

CHEMTREC Website Additional information about CHEMTREC operations and services can be obtained at its website, www.chemtrec.org. A section of the site is dedicated to emergency responders. This site was improved and updated recently to better meet the needs of users. Current information is provided on training opportunities, regulation changes, reports and other articles of interest for emergency responders. Helpful links to other hazmat sites are provided along with frequently asked questions.

> **Author's Note:** *During my first two visits to CHEMTREC, they had not yet computerized their response information. During my last visit in the early 2000s to do an article for* Firehouse Magazine, *they had not only computerized their chemical database but created their MSDS database as well.*

Planning for Hazardous Materials Incidents

Drexel Chemical Company Fire and Explosion

In the heat of summer 9:25 a.m., on July 5, 1979, about a year after the formation of the hazmat team, Memphis Fire and Hazmat faced a third alarm fire at the Drexel Chemical Company, 2387 Pennsylvania Avenue, near the intersection of Pennsylvania and Mallory streets. The metal building was 100 ft × 200 in. long. Numerous explosions occurred sending

Figure 2.44 Numerous explosions occurred sending 55-gallon drums flying through the air. (Courtesy Memphis Fire Department.)

55-gallon drums flying through the air (Figure 2.44). An 8,000-gallon tank of parathion, a pesticide, caught fire. In an interview with the Commercial Appeal maintenance mechanic David Trumble said of the initial explosion, "I heard one hell of an explosion and ran like hell." Another employee Robert Belden who was working in a nearby warehouse "heard an explosion. It was like a dull boom and I saw the doors of the warehouse coming off. Then it knocked me down, and the flash burned my hands in a couple of places and I went outside."

Water from the firefighting efforts spread the fire through a cache of methyl parathion which they poured into the Mississippi River because the fire could not be contained. Sulfuric acid was also dumped into the river as a toxic cloud spread beyond the plant, endangering the lives of area residents. Over 3,000 residents were evacuated including 400 who were taken by transit authority buses to an evacuation center set up at Whitehaven High School. Over 200 people visited hospitals complaining of bleary eyes, headaches and vomiting. Treatment was hampered by a lack of medical familiar with treatment of chemical exposures.

In 1984, Memphis local government took steps to help prevent another disaster like the Drexel Chemical fire. A hazardous materials advisory committee was formed with representatives from emergency services, government and industry who secured a $100,000 federal grant to create a hazardous materials plan. As a result of the plan and extensive training, the Chemical Manufacturers' Association declared in 1986 that Memphis was the American city best prepared to cope with a hazardous materials emergency (*Firehouse Magazine*).

The Critical Importance and Implementation of ICS and Incident Management

Introduction

While potential hazards facing communities are diverse, hazardous materials and terrorism are the primary focus of ICS and incident management in this volume. Communities may be large or small, but the threat is not limited to the size of a community (Figure 2.45). However, even though the potential threat is constant, the resources available to a community are not. Because major hazardous materials incidents are often beyond the capability of a single community regardless of size, it is important that all responders are able to function within an Incident Command System (ICS). Using a system allows for responders to better manage an incident and foster a successful outcome.

FEMA has developed a National Incident Management System (NIMS), which allows for responders and agencies across the spectrum to effectively work together under a common system. NIMS provides for common terminology, which allows a smoother meshing of agencies who may not have worked together before. NIMS integrates federal, state, and local government planning, preparation, and response capabilities. NIMS presents a single Incident Command System for use by government agencies nationwide, with a standard structure for the effective management of incidents of all types.

NIMS ICS system is based on the ICS developed by FIRESCOPE used in California Forest Service for decades. Public Information Officer (PIO)

Figure 2.45 Small communities are not immune from hazardous materials incidents. However, their resources are limited. (Courtesy Southeast Community College.)

is the only function of the system that differs from FIRESCOPE. All jurisdictions should be familiar with NIMS even if they only use the full blown version for Level III incidents. Because, in my opinion, it is too massive and not practical to be used on Level I and II incidents. Selected sections do have the ability to add to your capability for small scale incidents.

You need to use the Incident Command System and Incident Management system that are documented in your jurisdiction's emergency response plan for most incidents you will encounter. NIMS was developed following September 11, 2001, and meant generally for large-scale terrorist incidents in the beginning. It also works for large-scale hazmat incidents involving State and Federal resources. FEMA has the complete document in PDF format that can be downloaded from their website at no cost: https://www.fema.gov/national-incident-management-system.

Incident Command System (ICS)

Incident Command is divided into five Sections, Command, Operations, Planning, Logistics, and Finance (Figure 2.46). The Incident Command System is a flexible system that can be expanded or contracted depending on the size of the incident and resources available locally. For the vast majority of incidents, Command may be the only IC Section utilized. When command is the only Section, then the Incident Commander performs all of the other functions. The IC would also be the Safety Officer (SO), PIO, and Operations Chief. Other Sections of the ICS may not come into play formally. Some of the functions of those other sections may also be performed by the IC as part of the incident mitigation or termination, but informally.

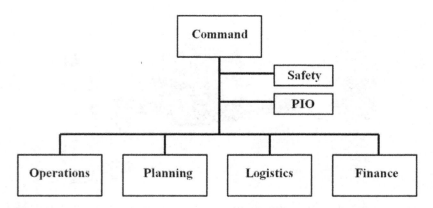

Figure 2.46 Incident Command is divided into five Sections, Command, Operations, Planning, Logistics and Finance.

Hazmatology Point: If your agency is not fully prepared and capable in terms of resources, equipment, and properly trained personnel to intervene, defensive or non-intervention strategies will likely be the preferred Operations option.

Command is the decision-making Section and responsible for overall incident safety and effective outcome (Figure 2.47). It can be single or unified command. All information from the other Sections comes through Command to assist the IC in making informed decisions about goals and mitigation actions. Once decisions are made at the Command Level, they are passed down to the other Sections for implementation. Operations Section is where all of the mitigation actions take place. Generally, there are several Branches to the Operations Section. Underneath the Operations Section, there may be a Hazmat Branch, Fire Control Branch or EMS Branch. Branches are added as an operational function is required to complete a portion of the Plan of Action.

Incident Commander Responsibilities
- Authority over incident.
- Incident Safety.
- Creates Command Post.
- Sets Priorities, Determines incident objectives and Strategies.
- Develops ICS Organizational Structure.
- Approves Incident Action Plan (IAP). An **incident action plan** (IAP) formally documents **incident** goals, operational period objectives, and the response strategy defined by **incident** command during response **planning.**

Figure 2.47 Command is the decision making Section and responsible for overall incident safety and effective outcome. (Courtesy Metro West Fire Protection District.)

- Coordinates Command and General Staff responsibilities.
- Approves Resource Requests.
- Confirms after action report completion and distribution.
- Authorizes media information releases.
- Authorizes Demobilization (Incident Termination).

Incident Management Team (IMT)

An IMT is a collective assembly of ICS-qualified persons including, Incident Commander, Command, and General Staff. Training and experience of team members coupled with formal response requirements and responsibilities are factors in determining type, or level, of the IMT.

Command Staff

Members are assigned to perform functions required to support the IC. These may include Interagency Liaison, Public Information Officer (PIO), and Safety Officer (SO).

General Staff

General Staff represents and is responsible for functional aspects of the Incident Command Structure. Staff positions include Operations, Planning, Logistics, and Finance/Administration Sections. If necessary, an Intelligence/Investigation Function may be established and function either operating under a staff section or as a standalone section.

General guidelines related to General Staff positions include the following:

- Only one person will be designated to lead each General Staff position.
- May be filled with qualified persons from any agency or jurisdiction.
- Members of the General Staff report directly to the Incident Commander. Any General Staff position not activated will become the responsibility of the IC.
- Deputy positions may be established for each of the General Staff positions. Deputies are individuals fully qualified to fill the primary position. They can be selected from other jurisdictions as appropriate. This fosters greater interagency coordination.
- General Staff members may exchange information with any person within the organization. Direction takes place through the chain of command. This is an important concept in ICS.
- General Staff positions should not be combined.

Public Information Officer (Figure 2.48)
- Determine, according to direction from the IC, and limits on information release.
- Develop accurate, accessible, and timely information for use in press/media briefings.
- Obtain IC's approval of news releases.
- Conduct periodic media briefings.
- Arrange for tours and other interviews or briefings that may be required.
- Monitor and forward media information that may be useful to incident planning.
- Maintain current information, summaries, and/or displays on the incident.
- Participate in planning meetings.

Liaison Officer
- Act as a point of contact for agency representatives.
- Maintain a list of assisting and cooperating agencies and agency representatives.
- Assist in setting up and coordinating interagency contacts.
- Monitor incident operations to identify current or potential inter organizational problems.
- Participate in planning meetings, providing current resource status, including limitations and capabilities of agency resources.
- Provide agency-specific demobilization information and requirements.

Figure 2.48 The Public Information Officer (PIO) is responsible for disseminating information from the IC to the media.

Operations Section Chief

This section chief is responsible for managing all tactical operations at an incident. The Incident Action Plan (IAP) provides the necessary guidance. The need to expand the Operations Section is generally dictated by the number of tactical resources involved and is influenced by span of control considerations.

Major responsibilities of the Operations Section Chief:

- Assure safety of tactical operations.
- Manage tactical operations.
- Develop the operations portion of the IAP.
- Supervise execution of operations portions of the IAP.
- Request additional resources to support tactical operations.
- Approve release of resources from active operational assignments.
- Make or approve expedient changes to the IAP.
- Maintain close contact with IC, subordinate Operations personnel, and other agencies involved in the incident.

Planning Section Chief

Provides planning services for the incident. Under the direction of Planning Section Chief, the Planning Section collects situation and resources status information, evaluates it, and processes the information for use in developing action plans. Dissemination of information can be in this form of the IAP, in formal briefings, or through map and status board displays.

Major responsibilities of the Operations Section Chief:

- Collect and manage all incident-relevant operational data.
- Supervise preparation of the IAP.
- Provide input to the IC and Operations in preparing the IAP.
- Incorporate Traffic, Medical, and Communications Plans and other supporting materials into the IAP.
- Conduct and facilitate planning meetings.
- Reassign personnel within the ICS organization.
- Compile and display incident status information.
- Establish information requirements and reporting schedules for units (e.g., Resources and Situation Units).
- Determine need for specialized resources.
- Assemble and disassemble Task Forces and Strike Teams (or law enforcement Resource Teams) not assigned to Operations.
- Establish specialized data collection systems as necessary (e.g., weather).

- Assemble information on alternative strategies.
- Provide periodic predictions on incident potential.
- Report significant changes in incident status.
- Oversee preparation of the Demobilization Plan.

Logistics Section Chief

The Logistics Section Chief provides all incident support needs with the exception of logistics support to air operations.

Major responsibilities of the Logistics Section Chief:

- Facilities.
- Transportation.
- Communications.
- Supplies.
- Equipment maintenance and fueling.
- Food services (for responders).
- Medical services (for responders).
- All off incident resources.

Finance/Administration Section Chief

The Finance/Administration Section Chief is responsible for managing all financial aspects of an incident. Not all incidents will require a Finance/Administration Section. Only when the involved agencies have a specific need for finance services will the Section be activated.

Major responsibilities of the Finance/Administration Chief:

- Manage all financial aspects of an incident.
- Provide financial and cost analysis information as requested.
- Ensure compensation and claims functions are being addressed relative to the incident.
- Gather pertinent information from briefings with responsible agencies.
- Develop an operating plan for the Finance/Administration Section and fill Section supply and support needs.
- Determine the need to set up and operate an incident commissary.
- Meet with assisting and cooperating agency representatives as needed.
- Maintain daily contact with agency(s) headquarters on finance matters.
- Ensure that personnel time records are completed accurately and transmitted to home agencies.
- Ensure that all obligation documents initiated at the incident are properly prepared and completed.

- Brief agency administrative personnel on all incident-related financial issues needing attention or follow-up.
- Provide input to the IAP.

Hazmat Incident Command

The National Incident Management System is great for large scale, what I like to call, Once In A Lifetime or Career Events. However, the reality of Hazmat Response is that those types of incidents rarely happen. What we can best do to prepare for a realistic hazmat response is to train and practice on incident command and incident management that will effectively deal with local incidents most likely to happen. This prediction of likely occurrences is based on type of chemicals present in a community, experience with incidents in the community and historical national incidents of the past. Research has shown that historically only a few chemicals over the years have killed and injured emergency responders. These include liquefied petroleum gases, anhydrous ammonia, chlorine, ammonium nitrate, and petroleum products such as gasoline and benzene. The United States Department of Transportation (DOT) has reported that almost half of all hazardous materials incidents involve petroleum products. Awareness of historical incidents and being able to field an effective incident command and incident management system can assure an effective and safe response.

Information presented here is designed to overview a fairly simple, effective incident command and incident management system that most any agency can implement for routine hazmat incidents. The material discussed here is just an overview and is meant for helping planners during plan development so they may understand the concepts they are using in their hazardous materials response plans.

Command

Collectively, Command is the Incident Commander and their staff. For example on an incident scene, an officer wants to undertake an initiative not previously known to the Incident Commander or Staff. That officer might say, I need to run this through Command before taking actions.

Organization and Function

Single Command

Generally, only one agency is involved in the command and control of an incident scene. A single incident commander is designated by the agency through their response plans to have overall management responsibility.

Unified Command

Applies the ICS when multiple jurisdictions are involved in an incident. ICs within the unified command make joint decisions and speak as one voice. Unity of command is maintained and each responder only answers to one boss. Each agency involved within the unified command has a designated commander for that organization who represents that agencies interests and resources within the unified command.

ICS Structure

Branch

ICS organizational level having functional responsibility for major segments of incident operations. The Branch level organizationally situated between Section and Groups under Operations and Section and Units in Logistics.

Division

Responsible for operations within a defined geographic area or with functional responsibility. The Division level is organizationally situated below the Branch.

Group

Established to divide the incident into functional areas of operations. Groups are composed of resources assembled to perform a special function not necessarily within a single geographical Division.

Unit

ICS organizational level having functional responsibility. Units are commonly used in incident planning, Logistics, or Finance/Administration sections and can be used in operations for some applications. Units are also found in some EOC organizations.

Other Commands and Considerations

Area Command

Is used when several incidents similar in nature are occurring in the same general area. Examples would be two or more hazardous materials incidents or fires or a terrorist incident with multiple related incident scenes.

Hazmat Incident Commander

First Arriving Unit Becomes the IC (OSHA 1910.120(q)(3)(i)

When dealing with hazardous materials releases, OSHA says "the most senior person responding to an emergency shall become the individual

Figure 2.49 OSHA says "the most senior person responding to an emergency shall become the individual in charge of a site specific ICS". (Courtesy Jacksonville, FL Fire Department.)

in charge of a site specific ICS" (Figure 2.49). All emergency responders and their communications shall be coordinated and controlled through the individual in charge of the ICS assisted by the senior official present for each agency. The senior official at an emergency response is the most senior official on the site who has the responsibility for controlling the operations at the site. In the fire service, initially it is the senior officer on the first due-piece of emergency responding apparatus to arrive on the incident scene.

As more senior officers arrive (i.e. Battalion Chief, fire chief, state law enforcement official, site coordinator, etc.), the position is passed up the line of authority which has been previously established.

> ***Authors Note:*** *Generally in the fire service, depending on local SOPs/ SOGs when the first fire company arrives at a scene, they have two options, take command, or pass command. This only involves the first arriving company. If nothing is obvious upon arrival at the scene, the company officer will pass command and further investigate. The next arriving company officer then must take command. However, if there is fire or heavy smoke showing, the company officer will assume command and direct operations as other companies arrive until relieved by a higher ranking officer. Additionally, in some departments, the higher ranking officer also has the option of taking or not taking command. If the higher ranking officer sees the situation is under control, they may let the officer in charge continue command of the incident. This allows for some on the job training for those officers who have aspirations of becoming a higher ranking office in their career.*

Safety Officer

Required (OSHA 1910.120(q)93)(vii)

The only other specific person mentioned and required in the OSHA hazardous materials regulations is the Safety Officer (Figure 2.50). "The individual in charge of the ICS shall designate a Safety Officer who is knowledgeable in the operations being implemented, at the emergency response site, with specific responsibility to identify and evaluate and to provide direction with the respect to the safety of operations of the emergency at hand."

> **Authors Note:** *In my opinion I believe the HM Safety Officer should be a technician-level trained person. After all, how would someone be able to determine the safety of hazmat operations without the knowledge and training of those who are conducting the operations.*

Safety Officer Power to Terminate Unsafe Operations (OSHA 1910.120(q)(3)(viii)

"When activities are judged by the Safety Officer to be immediately dangerous to life and health (IDLH), and/or to involve an imminent danger condition, the Safety Officer shall immediately inform the individual in charge of the ICS of any actions needed to be taken to correct these hazards at the emergency scene."

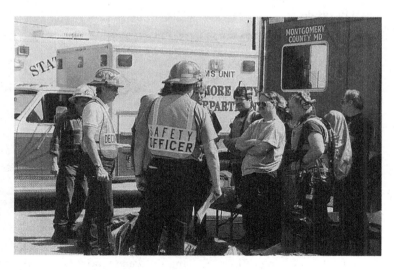

Figure 2.50 The only other specific person mentioned and required in the OSHA hazardous materials regulations is the Safety Officer.

Transfer of Command

Higher ranking officers arriving on scene who wish to take command must go through a formal process of transfer of commend (Figure 2.51). Transfer of command during an expanding incident is something to be expected and has no reflection on the current officer in charge. Transfer of command is an important part of the process of managing an incident. There are some general procedures that should be followed to insure the change of command is orderly and effective without interrupting the operations already in progress.

- The transfer of command should occur face to face if at all possible. The incoming IC should personally perform an assessment of the incident situation with the existing IC.

  Authors Note: Many times in my career I have heard a higher ranking officer on the air in service responding to the incident declaring they are assuming command. It is really not very effective to have an IC on scene and an in route IC taking command before they know anything about what is happening on the incident scene. Department operating procedures should prohibit such activity if at all possible.

- The incoming IC must be adequately briefed. This briefing must be by the current IC, and take place face to face if possible. The briefing must cover the following:
  Incident History (what has happened).
  Priorities and objectives.
  Current plan.

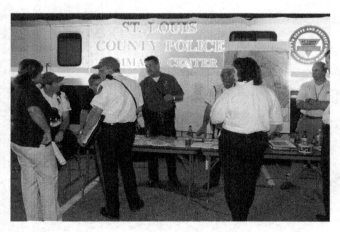

Figure 2.51 Higher ranking officers arriving on scene who wish to take command must go through a formal process of transfer of commend. (Courtesy St. Louis County, MO Hazmat.)

Resource assignments.
Incident organization.
Resources ordered or needed.
Facilities established.
Status of communications.
Any constraints of limitations.
Incident potential.
Delegated authority.

- After the incident briefing, the incoming IC should determine an appropriate time for Transfer of Command.
- At an appropriate time notice of a change in Incident Command should be made to:
Agency headquarters (through dispatch).
General Staff members (if designated).
Command staff members (if designated).
All incident personnel.
- The incoming IC may give the previous IC another assignment on the incident.
    There are several advantages to this:
The initial IC retains first-hand knowledge of the incident site.
This strategy allows the initial IC observe the progress of the incident, and to gain experience.

Essential Elements of ICS

Common Terminology

ICS utilizes common terminology so that all jurisdictions that respond to an incident will understand the system and be able to communicate effectively. Using common terminology helps to define organizational functions, incident facilities, resource descriptions and position titles.

Modular Organization

The incident command organizational structure develops in a top-down modular fashion that is based on the size and complexity of the incident, as well as the specifics of the hazard environment created by the incident (Figure 2.52).

Management by Objectives

Includes establishing over arching objectives; developing and issuing assignments, plans, procedures, and protocols; establishing specific, measurable objectives for various incident management functional activities; and directing efforts to attain the established objectives.

Hazardous Materials Command Structure

Figure 2.52 The incident command organizational structure develops in a top-down modular fashion that is based on the size and complexity of the incident.

Reliance on an IAP

IAPs provide a coherent means of communicating the overall incident objectives in the context of both operational and support activities.

Chain of Command

Chain of command refers to the orderly line of authority within the ranks of an incident management organization.

Unity of Command

Every individual has a dedicated supervisor to whom he or she reports at the scene of an incident (Figure 2.53). These principles clarify reporting relationships and eliminate the confusion caused by multiple, conflicting directives. Incident managers at all levels must be able to control the actions of all personnel under their supervision.

Unified Command

In incidents involving multiple jurisdictions, a single jurisdiction with multiagency involvement, or multiple jurisdictions with multiagency involvement will require a unified command structure. Unified command allows agencies with different legal, geographical, and functional authorities and responsibilities to work together effectively without affecting individual agency authority, responsibility, or accountability.

Figure 2.53 Every individual has a dedicated supervisor to whom he or she reports at the scene of an incident.

Manageable Span of Control

Span of control is key to effective and efficient incident management. Within ICS the span of control of any individual with incident management supervisory responsibility should range from 3 to 7 subordinates. Beyond this range, the ability of supervisors to effectively manage the personnel is diminished.

Pre-designated Incident Locations and Facilities

Various types of operational locations and support facilities are established in the vicinity of an incident to accomplish a variety of purposes. Typical pre-designated facilities include ICPs, Camps, Staging, Mass casualty triage areas and others as required.

Resource Management

Resource management includes processes for categorizing, ordering, dispatching, tracking and recovering resources. It also includes processes for reimbursement for resources as appropriate. Resources are defined as personnel, teams, equipment, supplies and facilities available or potentially available for assignment or allocation in support of incident management and emergency response activities.

Integrated Communications

Incident communications are facilitated through development and use of a common communications plan and interoperable processes and architectures.

Transfer of Command

The command function must be established clearly from the beginning of an incident. When command is transferred the process must include a briefing that captures all essential information for continuing safe and effective operations.

Accountability

Effective accountability at all jurisdictional levels and within individual functional areas during incident operations is essential. To that end the following principles must be adhered to.

Check In: All responders regardless of agency affiliation must report to receive an assignment in accordance with the procedures established by the IC.

IAP: Response operations must be directed and coordinated as outlined in the IAP.

Unity of Command: Each individual involved in incident operations will be assigned to only one supervisor.

Span of Control: Supervisors must be able to supervise and control their subordinates adequately, as well as communicate with and manage all resources under their supervision.

Resource Tracking: Supervisors must report and record resource status changes as they occur.

Deployment

Personnel and equipment should respond only when requested and when dispatched by an appropriate authority.

Command Post Location

Operations for any given incident are directed and coordinated from the Incident Command Post. The location of the command post should be such that it is convenient to the incident scene but in an area safe and

Figure 2.54 The location of the command post should be such that it is convenient to the incident scene but in an area safe and unlikely to need relocation during incident operations. (Courtesy Metro West Fire Protection District.)

unlikely to need relocation during incident operations. (Figure 2.54). At the very least the following criteria should be followed.

Uphill.
Up Wind.
In The Cold Zone.
Within Outer Perimeter.

Authors Note: The Miamisburg, Ohio phosphorous incident presented responders with a number of challenges in their efforts to manage the incident. It required extensive evacuations, changes in wind direction required moving an ideal command post location to another safer location. The derailment location and fires were in a somewhat remote location that made access to the incident scene complicated.

Case Study Miamisburg, Ohio Phosphorus Incident
Miamisburg, OH July 8, 1986, Derailment Phosphorus Fire

July 8, 1986, at 4:25 p.m. a southbound Baltimore and Ohio Railroad Company (B&O) freight train Southland Flyer (FLFR) derailed near Miamisburg, Ohio. The train was made up of a locomotive and 44 cars traveling over the Bear Creek Bridge. The locomotive and 27 cars made it over the trestle, remained coupled together and as the next 11 cars moved over the trestle they derailed. Only one car on the train was carrying a considered hazardous material, and it derailed, turning on its side, it was the phosphorus tank car. Near the phosphorus car were a tank car of animal fat, which was leaking into the creek and a box car of rolled paper, neither of which were involved in the fire. Phosphorus is air reactive and shipped under water to keep it from spontaneously combusting. Several holes developed in the tank during the impact of the derailment and the water started leaking off causing the phosphorus to be exposed to air and it ignited (Figure 2.55).

Fire Inspector Chuck Stockhauser was going to the west side of the river and a citizen stopped him and told him about the derailment. At 16:29, a General Alarm was sent out over pagers to call in all firefighters. Firefighter/Paramedic Steve Meadows, now a Battalion Chief, was off duty and came into town. He noticed the train had stopped and thought that was unusual. He saw Captain Sutton driving Engine 3, got in and heard there was a derailment. Captain Mike Sutton responded from the station with three firefighters. When they arrived on scene they began evacuations immediately of a few homes around the derailment site with the help of police officers and Fire Inspector Stockhauser.

Figure 2.55 Several holes developed in the tank during the impact of the derailment and the water started leaking off causing the Phosphorus to be exposed to air and it ignited. (Courtesy of Special Collections & Archives, Wright State University).

Captain Sutton checked with the railroad conductor and was told there was only one car containing hazardous materials and that was the one on fire. There were minor explosions occurring, which may have been the gas tanks of automobiles on the two car carriers that derailed. Captain Sutton's crew pulled lines and prepared to fight the fire. Mutual aid was requested from 30 area fire departments along with 107 medic units. Police agencies sent personal from Cincinnati, Dayton sent 25 police units along with other communities that sent help as well.

Meadows stayed with Engine 3 and ran the pump full throttle for 2 hours before he was relieved. Within two hours 3200 g.p.m. of water was being pumped through 4 master streams onto the fire. Water supply was an issue because hydrants were not close. Four-inch hose was used with inline pumpers to supply water. Firefighters in or near the phosphorus cloud wore SCBA, which was refilled by the air cascade system. Air truck had to be taken to West Carrolton to be filled as they were using a lot of air.

Responders went to cartridge respirators, supplied by the Emergency Management Agency (EMA) on the second day or so. Many restaurants brought food for the firefighters including steaks. Steve Meadows helped deliver the food. None of the food came from Miamisburg, because the town had been evacuated. There were no security issues, and there was no looting during the evacuation. Miamisburg looked like a ghost town.

Returning from the hospital from and EMS call, Andy Harp and his partner Leroy Kline heard the call on the radio and went to the scene in their medic unit. There was a 100' column of white smoke rising from the derailment site. They parked behind Rausch Contracting building for protection. At the time of the derailment Miamisburg had a population of 16,000. There were 23 career personnel and 14 reserves. Seven personnel along with the fire chief and fire inspector were on duty that day. They operated from 2 fire stations with 2 engines, 2 ladders, Air Unit, Brush Unit, 3 medics and two reserve engines. Robert Menker was the Fire Chief and Incident Commander (IC) at the time of the incident. He passed away in 2015.

When Chief Menker arrived on scene, he ordered an immediate evacuation of the west side of the Great Miami River. After evaluation of the incident status, he additionally ordered the northeast section of town evacuated as well. His amounted to about 10,000 people. Following consultation with other chiefs, it was decided to mount an attack on the fire utilizing water cannons and direct hose streams. The Train Master arrived at the command post and challenged the Chief's authority to handle the emergency and was subsequently escorted from the command post under threat of arrest. The winds changed in the evening of July 9th and coupled with a wind inversion that kept the phosphorus cloud close to the ground, prompted a second evacuation effecting approximately 30,000 people.

During the first and second days of the emergency numerous proposals for handling the phosphorous tank were suggested and evaluated.

Some of these included direct hose stream attack; plugging; water flooding of the interior of the tank car; foam application; burial; opening the manhole to allow air injection to accelerate the burn rate; and use of explosive demolition (it was suggested that the Air Force could use a Fighter from nearby Wright Patterson to shoot the tank car with a missile). Chief Menker in consultation with the city manager, decided to proceed with the suggestion of opening the manhole to accelerate the burn rate.

Initially the Command Post was established on a bridge upwind, south of the derailment (Figure 2.56). The bridge provided an excellent vantage point for the incident because the tracks ran right under the bridge with a view north of the tracks and the derailment site. However, the wind changed to the north and the command post had to be moved. Representatives from federal, state, and local agencies were at the IC. However, Chief Menker maintained absolute control over the incident. Radio communications was the biggest problem as many agencies were on separate frequencies, local mutual aid channels were overloaded with communications. So much so, that remote activation of evacuation sirens could not be accomplished when needed.

AT&T provided emergency phone bank at the command post. An emergency operations center was established at Station 1, where logistics and support were being provided. Firefighter Andy Sharp slept with his head on a rail the 1st night at the scene and at station 2, the second night on a roll of insulation. All but three firefighters lived in the city, so they could not go home to sleep. Those three went home to sleep on the third night. Firefighter ID's were not used on a daily basis that became an issue when firefighters were sent home to rest upon returning they did not have any way to prove they were firefighters. Police would not let them back in because of the evacuation in town and security set up to protect the evacuation area. As a result, identification badges were created for personnel operating at the incident.

Over the five day duration of the incident, crews were rotated in and out, millions of gallons of water were flowed onto the fire. By the time the emergency was declared under control, local hospitals received 569 persons with nonfatal injuries, including 13 emergency response personnel. Of these, 27 were hospitalized due to their injuries. The cost, excluding costs to evacuees, community disruption or business interruptions was estimated to be over 3.5 million U.S. dollars.

Safety

Safety Officer

The Safety Officer is responsible for all safety at an incident scene. The Safety Officer has the authority to alter, suspend or terminate any

Figure 2.56 Initially the Command Post was established on a bridge upwind, south of the derailment. (Courtesy Miamisburg, OH Fire Department.)

operation in an Immediately Dangerous to Life and Health (IDLH) atmosphere when he or she deems it necessary to protect the safety of any responder. Clearly, the person needs to be one of the most knowledgeable and competent individuals at the operation.

Responsibilities of the Safety Officer Include:

Safety Briefing
- Normally performed by the Assistant Safety Officer Hazmat. Ensures that proper personal protective equipment is used and decontamination is properly setup and ready.
- Conduct a pre-entry briefing.
- Provide for emergency medical services and first aid. OSHA requires EMS standing by. More than one unit is required if victims are involved. Treatment for heat and cold stress, injuries or chemical exposure may be needed.

Agency Liaisons

A Liaison Officer is activated when two or more jurisdictions or agencies are present at a hazardous materials incident. That person will coordinate interactions between agencies and the IC in order to successfully incorporate their resources into a particular incident.

Operations

Operations is where all physical activity takes place in order to mitigate an incident. Operations encompasses such things as fire suppression, EMS, site security, hazardous materials and others. For our purposes in this volume, we will focus on the hazardous materials group, HM Group (Figure 2.57).

Research

Determining the type of hazardous material present and the hazards are among the first things that need to be accomplished at an incident scene. That may be conducted by first responders before the hazmat team arrives. Once research duties are assigned, personnel also need to verify, identify and hazard information, determine appropriate PPE.

Decontamination (Decon)

If contamination has occurred or anticipated, then a decon system needs to be set up prior to entry into the Hot Zone by technician-level personnel (Figure 2.58). If the physical state of a hazardous materials is a gas, then minimal decontamination is necessary and does not require a full blown decon line. Some jurisdictions employ dry decontamination. Decontamination is defined as a chemical and/or physical process used to remove and prevent the spread of contaminants able to cause harm to life, health, and the environment.

Hazmat Group

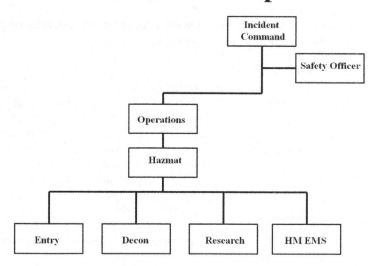

Figure 2.57 The focus of this volume under operations is the Hazardous Materials Group.

Figure 2.58 If contamination has occurred or anticipated, then a decon system needs to be set up prior to entry into the Hot Zone by technician level personnel.

Types of Contamination: Product residue on objects or persons.

Direct: Someone comes in contact with a contaminated surface and becomes contaminated.

Cross: Someone comes in contact with a contaminated person or piece of equipment that has not been properly deconed.

Forms of Decontamination

Emergency: First responder personnel can perform emergency decontamination from outside the Hot Zone from a position of up wind and uphill to get contaminants off of victims. This can be accomplished with equipment carried on most fire engines (Figure 2.59).

Mass: Mass decontamination may be performed by first responders to remove contaminants from victims that may cause serious harm without emergency decon. The first step in utilizing mass decon is to remove all outer clothing and properly bag it. Mass decon can be done again with equipment normally carried on fire engines. It can also be a function of hazmat teams when they arrive on scene with the use of mass decon trailers or tents. Consider privacy, hypothermia, and personnel showers.

Figure 2.59 First responder personnel can perform emergency decontamination, which can be accomplished with equipment carried on most fire engines.

Technical: Technical decon is performed by hazmat team members and is primarily for entry personnel (Figure 2.60), but is also used for non-ambulatory victims. Additional corridor(s), beyond responder corridors need to be set up for contaminated victims. Different needs for exist for ambulatory and non-ambulatory. (Figure 2.61)

Types of Decontamination

Gross: Gross decontamination is conducted with emergency decon or it is the first station of a technical decontamination line. In terms of technical decon it can be accomplished with a decon shower or hose line.

Secondary: Secondary decontamination occurs with a number of wash and rinse stations in the remainder of the technical decon line. The number of stations is determined by the extent of the contamination and the dangers of the contaminant.

Tertiary: Tertiary decontamination is removed at a medical facility and usually involves contamination of wounds and may require debriding.

Figure 2.60 Technical decon is performed by hazmat team members and is primarily for entry personnel.

Figure 2.61 Additional corridor(s), beyond responder corridors need to be set up for contaminated victims.

Methods of Decontamination

Wet or Dry: Wet decontamination involves the use of water or soap and water to remove the material from response personnel or victims. When Dry decontamination involves the use of fans or chemical wipes that remove the material from chemical suits.

Emulsification: Using a dry material to soak up the liquid contamination into the material and the material and liquid contamination are then removed.

Chemical reaction: Using a chemical that reacts with another chemical reducing the contaminant to less harmful products that can usually be washed off with water.

Disinfection: Generally used with biological contamination, disinfectants kill the biological material and then can be washed off with soap and water.

Removal: Simply removing the physical material usually by mechanical means.

Dilution: Usually using water, the water reduces the concentration of the contaminant to the point it is no longer harmful.

Absorption: Apply some type of dry material that absorbs the liquid contaminant into itself and then removed mechanically from the contaminated person or equipment.

Disposal: All contaminated materials including decontamination water must be properly disposed of according to local health department or environmental protection procedures.

Note: All decontamination using chemicals, chemical reactions, or water reactions are conducted only on equipment, not personnel or victims.

Contaminated Victims and Runoff Control

Life safety takes precedence over containment of runoff. US EPA guidance allows for non-collection of runoff for chemically contaminated victims when it is life and death situation. When practical, containment should be established.

Protected Personnel Following PPE Removal

Gross

Removal of enough contaminant to safely doff equipment. Residual contamination can be anticipated. The PPE will be required to be bagged and cleaned or disposed of.

Secondary

PPE: Decontamination occurs after the incident and may involve cleaning and testing of PPE to assure integrity following use.

Personnel:Should be minimal issue, at most requiring hand and face wash or complete shower following doffing of PPE.

Mass Decontamination Equipment

Trailers (Figure 2.62)
Tents (Figure 2.63)

Emergency Decontamination Equipment

Initial hand fog line fully stretched, aimed upward to provide focal point for victims. This included emergency responders who may have become contaminated. Helps protect apparatus and possibly stop mass exodus. Can be accomplished with equipment normally on firefighting apparatus. For larger groups, two fire engines side by side with discharge ports facing each can be equipped with fog nozzles put on wide angle to have contaminated victims walk through the spray (Figure 2.64). Aerial apparatus can also be deployed with the nozzle on wide angle fog and flowing down like a giant shower and have contaminated victims walk through.

Figure 2.62 Decontamination trailers for mass decon are available in many large jurisdictions.

Figure 2.63 Decontamination tents are portable and easy to assemble anywhere at an incident site. (Courtesy University of Maryland.)

Figure 2.64 For larger groups, two fire engines side by side with discharge ports facing each can be equipped with fog nozzles put on wide angle to have contaminated victims walk through the spray.

Decontamination Solutions

Only decontamination solution routinely used on personnel and victims is water or soap and water. Other solutions damage turnouts, SCBA's, other equipment (especially neutralizing and bleach solutions).

Nothing other than water or soap and water should ever be used on victims (neutralizers, bleach, solvents have caused significant health effects on victims).

Victim Transportation to Medical Facility

Victim and vehicle packaging should not be an issue for deconed victims.

Bloodborne pathogen PPE will provide protection. Notify medical facility early.

Treatment specific to contaminant. Airlift generally not an option.

Entry:

PPE: Protective equipment and respiratory protection are worn to protect personnel from common hazmat hazards. Equipment worn is based upon potential hazards in the Hot Zone (Figure 2.65).

Figure 2.65 Equipment worn is based upon potential hazards in the Hot Zone.

Hot Zone Hazards can be any of the following:

Thermal: Can be a hot or cold material hazard. Hot surfaces, molten hazardous materials, fire or radiant heat can be present in a Hot Zone and cause melting, burning, and damage to PPE. Cold surfaces, cryogenic (extremely cold) liquids can cause PPE to become brittle and fail or shatter into pieces.

Radiation: Radiation can only be detected with meters. Some PPE can be penetrated by radiation causing harm to personnel. Radiation exposure must be limited and is generally acceptable for rescuing victims.

Asphyxiation: There are two basic types of asphyxiation, chemical and simple, which is a general lack of oxygen in the air to support life. Chemical asphyxiation occurs when a chemical enters the body and prevents the body from using available oxygen. There is plenty of oxygen available, but the body cannot gain access to it.

Chemical: Chemical hazards are corrosivity, toxicity, reactions with other chemicals and water.

Etiologic: Disease causing agents also sometimes referred to as biological materials. These are living organisms that can cause disease in humans, animals, and plants.

Mechanical: Mechanical hazards are anything that can physically harm responders. These can be falls, sharp objects, trip hazards, electrical hazards, and others.

PPE must be designed to protect against the specific hazard or hazards to be encountered by emergency responders entering the Hot Zone (Figure 2.66).

Primary hazard
    Most severe hazard
    May be more than one
Secondary hazard
    Not as severe as the primary
    Must be addressed as well
Buddy System: Responders entering the Hot Zone must do so in teams of two (Figure 2.67).

Figure 2.66 PPE must be designed to protect against the specific hazard or hazards to be encountered by emergency responders entering the Hot Zone. (Courtesy Philadelphia Fire Department.)

Figure 2.67 Responders entering the hot zone must do so in teams of two, called the Buddy System.

Back-up Personnel: Back up personnel, dressed out but not yet on air must be standing by to make quick entry to assist the entry team if they get into trouble.

Case Study: Allegany County, PA Special Intervention Team

Introduction

If you were to look at the meters and air monitors that were available to hazardous materials responders in the 1980s during the dawn of hazmat response and compared them to today's world of 4 gas meters with integral PIDs, infrared spectrometers and even truly portable gas chromatographs, the advances in just 30 or so years are astounding. Much of the same can be said for the PPE and information technology available to responders. In spite of that, the hazmat community is still teaching new technicians to decontaminate in the same way our predecessors did 30 years ago.

After nearly 20 years in the hazardous materials business, we decided there has to be a better way to decontaminate people in the field. The impetus for this thought was an episode that played out during a National Pro Board Hazardous Material Technician practical exam. One of the evaluators wanted to fail the entire team on the decon station because some of the simulated contaminant used was found under the SCBA of the contaminated responder being cleaned. It was then that we first realized that we had gotten the whole concept of decon all wrong. Two years of deliberating, experimenting, and trials have led to the method outlined in this paper. A method that is more efficient in time, water, and manpower without risking the safety of the responder. Our colleagues and us poured a lot of our life into this idea and have accepted and evaluated much criticism, constructive or otherwise and the product is what we believe decon should and will look in the coming years.

False Assumptions

The first obstacle to overcome was a number of false assumptions about our current decontamination practices, which are loosely based on the EPA decon methods. For instance, many people start with the assumption that decon means completely cleaning any contamination from a responder at a technical decon line in the field. Once you start attempting to do this in a controlled training setting, it quickly becomes self-evident that totally cleaning someone is a fool's errand that is nearly impossible and very time consuming. Another close assumption is that someone's PPE *needs* to be completely clean. On the contrary, we propose that the outer PPE only needs to be clean enough. Clean enough for the responder and an assistant to safely remove the PPE from the responder without getting it on him. Once clear of the PPE without further spread of contamination, it doesn't matter if the PPE is completely clean or not, because it will simply be packaged for disposal. Another of the false assumptions we commonly live by is that everyone in the Hot Zone needs decontaminated.

Again, after applying much thought and good science, it is easy to see that this is just not the case. For instance, if responders were in a cloud of gas or vapor, such as chlorine or ammonia, what exactly is it that we would be decontaminating? In the case of liquids, if there is a puddle on the ground and it is obvious no one has been close to it, let alone stepping in it, do they need deconed? And lastly, if someone does step in a puddle of diesel fuel, would you really need to decontaminate their whole body or just their feet? As we dissected some of the currently accepted methods of decon, more and more sacred truths began to unravel under the scrutiny of logic and science.

Why Would We Want to Fix Something That Already Works?

Decontamination as it is popularly practiced in the country today is a resource intensive operation. When examined closely, is it really providing value commensurate with the resources it exhausts? For instance, if you set up a basic 3-step decon, it can involve anywhere from 3 to 7 people, gallons of water per person deconned and a plethora of hardware store "junk" including buckets, tarps, hoses, brushes, pools, and maybe even a portable shower. In addition to the amount of resources consumed, more taxing can be the time involved. Often, decon lines at larger incidents can be seen backing up and low air alarms can be heard ringing as decon teams race to move responders through the line in time to get everyone safely out of their PPE. Our method of decon both dramatically reduces the resources used and the amount of time it takes to decon responders. Using this method properly, a responder can be decontaminated and out of their PPE in under 3 minutes all while using only about 1 gallon of water per responder.

So How Does It Work?

The Common Sense Decon method works in 3 basic parts. The first is a decision-making flow chart that allows the decon team leader to quickly make important decisions for providing the most effective decontamination. This flow chart directs the decon decision maker in selecting whether or not to perform decon at all, whether a wet or dry method should be used, and when to send someone to doff their PPE.

The second part is the actual method for removing contamination. For situations involving dry solids, such as dust and powders, a HEPA vacuum is used as the primary means of removal. For wet solids or liquids, with consistencies from thin oils to heavy greases or sludges, a lower pressure electric pressure washer is used. In either case, only enough contamination is removed to allow the responder to safely remove their PPE. To spend time removing any more contamination is simply wasteful. Our experiments show that cleaning the hands and the zipper areas of suits or turnout gear is the most important, but on an individual basis, a determination can be made as to whether boots or heads may also need to be cleaned. The actual methods recommended are detailed in the section entitled "Removing Contamination".

The third part of the Common Sense Decon method involves the doffing of PPE. The basic method for chemical protective clothing is to start by removing any tape from around the gloves (for level B and C) and then opening the zipper. The area around the zipper should be clean from the previous station. Next, move to the top (after removing SCBA when in Level B) and begin rolling the inside out and down and over the shoulders trapping any remaining contamination inside as the suit is rolled down. As the suit is rolled to the shoulders, have the responder make a fist to capture their gloves and pull each sleeve inside out, letting go of the gloves after the sleeve is inside out, but in time to allow the glove to remain in the sleeve. Then, continue rolling the suit down to the feet and over the boots, leaving the rolled up suit around the boots, like turn-out pants pulled down around fire boots.

From here, the responder can step out of the boots and onto a clean adsorbent pad (or other clean surface). The decon team member assisting with decon can lift the boots and rolled down suit as a single unit and throw it into a waste drum. In another possible variation, before doffing, the responder can actually step unto a rolled down 55 or 85 gallon drum liner (Trash Bag) so that when they step out of their boots, the decon team member can simply close the bag, tie it, and then lift the whole bag into a waste disposal drum.

It is that simple. In testing, it proved consistently to work in a safe and effective manner and was able to push people through the decon line in about 2 minutes. In all tests, responders were able to remove their PPE with assistance from a decon team member without spreading contamination to themselves. The bottle neck in the system is the doffing station which accounts for about 75% of the time. The pressure washer can clean a responder in about 30 seconds using the procedure outlined in the "Removing Contamination" section. The system can be effectively run with just a washer and a PPE doffer, but to overcome the bottle neck caused by the doffing, we recommend one washer with two separate doffing stations accepting people alternately for doffing of PPE.

Removing Contamination

The saying, "The devil is in the details" is as relevant here as anywhere. To effectively remove contamination, there are several details that come into play. The first is the use of a pressure washer. After hearing all of the possible cons to this plan that could be dreamed up, we just decided to try it. We conducted experimentation involving turn out gear and chemical protective clothing. As contaminants we used everything from peanut butter to axle grease. Those experiments set us straight on a number of details. First, a pressure washer can be safely used on a person if the pressure is fixed low enough and the spray pattern is also permanently fixed at an acceptable angle.

We decided on an electric pressure washer at around 1200 psi which had interchangeable tips (hard to find on electric models) as opposed to the adjustable single tip. We decided that by selecting a tip with an appropriate angle such as 30° we wouldn't have to worry about someone mistakenly using a thin concentrated stream, which will damage suits and possibly injure people. In addition, we looked for a model that could induct soap right from an onboard tank without switching wands or nozzles.

The key to successfully and safely pressure washing a person whether in turn-out gear or chemical protective clothing is the method. This is one of the areas where specific training matters. The method we developed is not what we are traditionally trained to do. This method starts by pressure washing the area around the zipper (about 6″ on each side). The pressure washer should hold the gun high and point it at a downward angle. Starting at the top of the zipper, the pressure washer should work his way down the zipper in short up and down strokes, pushing any visible contamination downward. Once complete, a decision must be made. If the contaminated responder is in Level A, continue by pressure washing the hands and possibly the boots if heavy visible contamination exists there.

If the responder is wearing anything besides Level A, have them turn around and face rearward, extending their hands out in front of them. This allows for two critical things. If the suit is worn in the traditional way with the sleeves over top of the gloves, it allows the pressure washer to push contamination down the glove toward the finger and off the end without blowing it up under the sleeve, which could happen easily with a pressure washer despite the presence of tape. The other advantage of the deconee standing backward is that material can be blown off of the head if necessary without forcing it between the face piece and the hood. This was especially important on our team since we don't tape around face pieces. Our belief is that if you need to tape around a face piece, you need a different suit, which has led to the integration of encapsulated CPF3 or similar suits for some Level B work where heavy contamination, especially around the head, is likely.

Pressure washers will fling material and make a cloud of mist, so we decided we would need a backstop to collect the water and material that was coming off our deconee. The solution that we chose after a great deal of trial and error was a three-sided structure built from PVC pipe with a removable plastic cover. The backstop was built with commonly available materials from the local home improvement store and the plastic covers can be made for a relatively low cost and are easily removable for disposal. Although this problem could be solved an infinite number of ways by teams wishing to use the method, we chose a method that allowed for quick assembly by even those with little training and allowed for easy storage on our apparatus.

Another paradigm shift was our approach to collecting the decon water. In most cases...we are not. Looking at the responses by the 5 teams in our county over the last 10 years we determined that we just are not getting that contaminated.

Using our method is creating very little waste water and in every case where a wet decon was used in the past, environmental clean-up contractors were later called in to remediate the scene. At about 1 gallon per person deconed, we decided that the simplest answer was to just let the run-off water lie on the ground. The area contaminated by the water will be insignificant and can be remediated by the clean-up contractor if necessary. We also understand that there will be exceptions to this, in which case we will still carry one collapsible decon pool for water collection, although we don't anticipate needing it.

It is likely that few people would argue the need to improve the current decontamination systems in use within the world of hazardous materials response. The current methods are resource intensive including manpower, equipment, and training time. Worst, all of the resource drain typically leads to results that are marginal at best. Our Common Sense Decon method reduces the resource load while improving the efficiency of the decon process. The year 2014 will be the field testing period for this method within our team following months of development and experimentation. We hope this method will gain widespread acceptance and give teams everywhere a viable alternative to what they are currently doing or at least point them in the direction for making their own needed improvements to the system. When looking at the advances that have been made in the technology we use today, it becomes easy to see the need for updating decontamination processes as well so that all parts of our response operations remain progressive for the safety of the public and our responders.

Monitoring

The purpose of air monitoring is to identify and quantify airborne contaminants in order to determine the level of public protection needed (Figure 2.68). Monitoring is also conducted to determine if an airborne contaminant may affect the population in general in the vicinity of a release. Confined spaces are required to be monitored prior to personnel entering to perform a rescue to make sure the responders have the appropriate PPE and respiratory protection.

Figure 2.68 The purpose of air monitoring is to identify and quantify airborne contaminants in order to determine the level public protection needed.

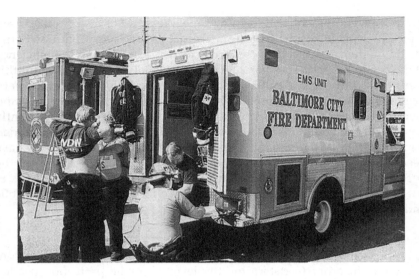

Figure 2.69 OSHA requires EMS standing by. More than one unit is required if victims are involved. Treatment for heat and cold stress, injuries or chemical exposure may be needed.

Hazmat EMS

OSHA requires EMS standing by. More than one unit is required if victims are involved.

Treatment for heat and cold stress, injuries or chemical exposure may be needed (Figure 2.69).

Requesting Additional Resources

Request Assistance

Resources vary tremendously depending on the locality and state having jurisdiction. Must rely on the Local Emergency Planning Committee (LEPC) and the State Emergency Response Commission (SERC) plan and agency SOG's for guidance on obtaining additional resources. In most cases dispatch or emergency management may actually notify and obtain additional resources.

Federal On-Scene Coordinator

Is a pre-designated representative from EPA or U.S. Coast Guard. Represents federal interests.

Can provide tremendous technical and coordination support.

Private Sector Assistance

CHEMTREC – Sponsored by American Chemical Council (formerly CMA).
Shippers.
Manufacturers.
Industrial mutual-aid groups, examples:
    Baltimore – SBIMAP South Baltimore Industrial Mutual Aid Plan.
    Houston Area – Channel Industries Mutual Aid.

Establish Communications

As needed, assign Communications Unit Leader. Augment communications with additional assets (mobile com center, etc.).

Joint Information Center (JIC)

The Joint Information Center (JIC) is a facility that houses Joint Information Systems (JIS) operations, where personnel with public information responsibilities perform essential information and public affairs functions. Depending on the needs of the incident, an incident-specific JIC may be established at an on-scene location in coordination with local, state, and federal agencies, or at the national level if the situation warrants.

Public Information Officer (PIO)

The PIO prepares public information releases for Incident Commander, Unified Command, EOC director, or MAC Group clearance. This helps ensure consistent messages, avoid release of conflicting information, and prevent adverse impact on operations.

> **Hazmatology Point:** *More firefighters have been killed fighting fires with ammonium nitrate than any other hazardous material. If the fires did not occur, we could save firefighter lives. Some important lessons can be learned from the West, Texas ammonium nitrate explosion. One of the issues noted by the Chemical Safety Board was the lack of use of an Incident Command System.*

Case Study: West, Texas Ammonium Nitrate Explosion

On April 17, 2013, a fire and subsequent explosion occurred at the West Fertilizer Company in West, Texas. Firefighters from the West Volunteer Fire Department were fighting a fire at the facility when the explosion occurred, approximately 20 minutes after they started fighting the fire

Figure 2.70 Firefighters from the West Volunteer Fire Department were fighting a fire at the facility when the explosion occurred. (Courtesy Chemical Safety Board.)

(Figure 2.70). Ammonium nitrate was located in a bin inside a seed and fertilizer building on the property. The explosion registered 2.1 on a seismograph reading from Hockley, Texas, 142 miles away. Fifteen people, mostly emergency responders, were killed, over 260 were injured and 150 buildings sustained damage. Victims included 5 volunteers with the West Volunteer Fire Department, 4 volunteers from neighboring volunteer departments that were attending an (EMS) class in West and went to the scent to help, one career captain off duty went to the scene to offer help, 2 civilians that went to help the West volunteers and 3 residents who lived around the facility. Investigators confirmed that ammonium nitrate was the source of the explosion. According to the United States Environmental Protection Agency (EPA), there was a report of 240 tons of ammonium nitrate on the site in 2012. According to the Department of Homeland Security, the company had not disclosed to them their ammonium nitrate stock. Federal law requires that the DHS be notified whenever anyone has more than 1 ton of ammonium nitrate on hand, or 400 pounds if the ammonium nitrate is combined with combustible material.

Ammonium Nitrate Hazards

Under normal conditions, pure solid ammonium nitrate is a stable material; it usually is not sensitive to mild shock or other typical sources of detonation (such as sparks or friction). However, AN exhibits three main hazards in fire situations:

- Uncontrollable fire.
- Decomposition with the formation of toxic gases.

Key Contributing Factors to Emergency Responders' Fatality

CSB identified the following seven key factors that contributed to the fatalities of firefighters and other emergency responders in West:

1. Lack of incident command system.
2. Lack of established incident management system.
3. Lack of hazardous materials (HAZMAT) and dangerous goods training.
4. Lack of knowledge and understanding of the detonation hazards of FGAN.
5. Lack of situational awareness and risk assessment knowledge on the scene of an FGAN-related fire.
6. Lack of pre-incident planning at the WFC facility.
7. Limited and conflicting technical guidance on AN.

Lack of Incident Command System

CSB found that none of the responding emergency response personnel trained and certified in the National Incident Management System (NIMS) process formally assumed the position of Incident Commander (IC) who would have been responsible for conducting and coordinating an incident command system (ICS). Senior emergency response personnel at the WVFD arrived at the scene of the WFC incident at different times and did not delegate an IC to be in charge of the incident. Also, there was no record that arriving firefighters conducted an initial incident size-up or risk assessment to determine initial actions (offensive or defensive) that would be most suitable in responding to the incident based on the situation and available resources without putting emergency personnel at risk.

Despite multiple responders having ICS training, none of them reportedly established command or took control of the fire ground. On the basis of a review of radio communications and interviews with surviving firefighters, CSB found no clear messaging or discussion among the responding firefighters on who should assume the role of the designated IC.

Without a delegated IC officially taking control of the fire ground operations, no ICS was established. Consequently, no senior emergency response personnel or IC was responsible for coordinating the various response activities carried out by individual firefighters on the scene. The

West fire chief arrived on scene at about 7:41 pm and did not critically assess the conditions on the ground before the explosion 10 minutes later, at about 7:51 pm. The fire chief and assistant chief provided support and advisory functions but did not actively engage in fire ground function or take control of the fire ground; no record indicated that the West fire chief took command of the incident upon his arrival. Without direction to the contrary, the firefighters immediately took offensive action against the flames coming from the doors on north end of the east side of the structure. CSB interviews with surviving firefighters indicated that before the arrival of the fire chief, the other senior firefighters who had reached the incident scene about six minutes earlier had not delegated senior personnel with the training and expertise needed to formally assume responsibility as the IC. The firefighters had not reached a conclusion about how to establish a best approach and how to respond to the fire when the explosion occurred. Despite being trained for the ICS and NIMS process, none of the certified firefighters had prior practical experience in establishing incident command or coordinating and maintaining control of any previous emergency that merited the same approach as an FGAN-related fire scene.

Lack of Established Incident Management System

CSB found that the emergency response personnel who responded to the WFC incident did not take time to set up, implement, and coordinate an effective incident management system plan that would have ensured evacuation of the nearby residents. Because no formal IC was in charge of the incident, none of the firefighters took responsibility for formally establishing and coordinating an effective incident management system.

Firefighter Training

Firefighters must cope with extraordinary situations and circumstances that threaten their personal safety. To improve execution and reduce the threat of injury or loss of life, it is vital for both volunteer and career firefighters to receive thorough training and information supporting effective decision making. CSB's investigation of the WFC incident revealed that no standardized training requirement applies to volunteer firefighters across the nation.

Nationally, CSB found that the curriculum used for HAZMAT training does not fully address the hazards and severity of FGAN-related fires and explosions. A review of the U.S. Fire Administration (USFA) National Fire Academy HAZMAT field course outlines confirmed that they place little emphasis on emergency response to storage sites containing dangerous reactive chemicals and oxidizers such as FGAN.

CSB concludes that the current training resources at the local, state, and federal levels do not provide sufficient information for firefighters to understand the hazards of FGAN. It is therefore essential for firefighter and emergency response training institutions to collaborate with fire departments to develop and implement a realistic process for ensuring that hazard response knowledge, once attained, does not become unused and obsolete.

Although many firefighter training courses provide overviews of initial fire scene size-up, assessment, incident planning, and execution, CSB found that none of the firefighter HAZMAT field training courses provide sufficient information on firefighter situational awareness and risk assessment that could help them make informed decisions while at the fire scene. The firefighters who initially responded to WFC did not have the tools to effectively perform the situational awareness and risk assessment that would have enabled them to make an informed decision to not fight the fire. Situational awareness in firefighting involves the capability to "read" the scene of a fire or emergency, including changes in the behavior of a fire. Effective situational awareness supports prompt decision making to either evacuate the scene of a fire or continue fighting the fire by taking a defensive or offensive stance. Chapter 4 of NFPA 472 (*Standard for Competence of Responders to Hazardous Materials/Weapons of Mass Destruction Incidents*), provides guidance on situational awareness competencies for responder-level personnel.

In fires involving HAZMAT, it is not always possible for firefighters to obtain needed information before acting, but they might be able to characterize a HAZMAT incident based on initial information acquired from the emergency call center and dispatcher; emergency response manuals and guides; knowledge base on the response area; and visual, auditory, and olfactory (odorous) clues. In some cases, the fire department's standard operating procedures (SOPs) and the level of training of the emergency response crew might be insufficient to respond at the incident scene to changing events and scenarios that were not planned for or anticipated – hence, the need for effective training on situational awareness and risk assessment.

The fire department did not have a formal pre-incident planning program for FGAN at WFC. Firefighters responding to the incident were aware of the risks associated with anhydrous ammonia leaking from the tanks and that it could form a toxic flammable cloud that could leave the facility, drift into nearby homes, and potentially explode. Although some responding firefighters knew that FGAN was onsite, they did not anticipate a possible FGAN explosion. Some of the West fire department officials reported that they were aware of the chemicals routinely stored at the WFC, but there was never any formal training to prepare for a fire or chemical emergency. Effective site-specific pre-incident planning for emergency responders is essential to guide initial and subsequent actions while responders are at an emergency. Onsite pre-incident planning might have identified the possible FGAN explosion hazard. CSB did not find evidence of regularly scheduled training exercises to ensure that the WVFD conducted incident pre-planning and facility tours to address fire safety and chemicals onsite.

Most firefighting apparatuses have a copy of the ERG. After the WFC incident, NIOSH investigators found copies of the 2012 ERG in the glove boxes of some of the damaged fire equipment and apparatuses. However, CSB does not have any evidence that indicates whether the West firefighters consulted the ERG on the night of the explosion. The ERG is especially useful in situations when the relevant SDS is not readily available to firefighters. The ERG gives direction (based on DOT Hazard Classification Criteria) on response to HAZMAT and dangerous goods emergencies during transportation.

Lessons Learned

On May 29, 2014, at around 5:45 pm, a fire involving FGAN occurred at the East Texas Ag Supply facility in downtown Athens, Texas. Emergency dispatchers and the Athens Police Department promptly notified firefighters from the Athens Fire Department (AFD). Emergency response units from the AFD arrived on the scene of the fire at 5:50 pm and found fire and smoke coming from the northwest end of the 3,500-square-foot East Texas Ag Supply facility. This facility was built with masonry bricks and combustible wooden structures, similar to construction at the WFC facility. The AFD chief arrived about 2 minutes after the first responding units were dispatched to the site of the incident, and he found that the fire had self-ventilated at the northwest end. On the basis of his observation of the enormous scope of the fire and the possibility of detonation of FGAN in the engulfed building, the fire chief promptly decided to let the East Texas Ag Supply facility burn to the ground instead of attempting to fight the fire. He ordered his firefighters to retreat from the scene and began an extensive evacuation of the downtown Athens, Texas, area.

Key Findings

The presence of combustible materials used for construction of the facility and the fertilizer grade ammonium nitrate (FGAN) storage bins, in addition to the West Fertilizer Company (WFC) practice of storing combustibles near the FGAN pile, contributed to the progression and intensity of the fire and likely resulted in the detonation. The WFC facility did not have a fire detection system to alert emergency responders or an automatic sprinkler system to extinguish the fire at an earlier stage of the incident.

The West Volunteer Fire Department (WVFD) did not conduct pre-incident planning or response training at the WFC facility to address FGAN-related incidents because was no such regulatory requirement. Thus, the firefighters who responded to the WFC fire did not have sufficient information to make an informed decision on how best to respond to the fire at the fertilizer facility.

Federal and state of Texas curriculum manuals used for hazardous materials (HAZMAT) training and certification of firefighters placed little emphasis on emergency response to storage sites containing FGAN.

Lessons learned from previous FGAN-related fires and explosions were not shared with volunteer fire departments, including the WVFD. If previous lessons learned had been applied in West, the firefighters and emergency personnel who responded to the incident might have better understood the risks associated with FGAN-related fire.

CSB's analysis of the emergency response concludes that the West Volunteer Fire Department did not conduct pre-incident planning or response training at WFC, was likely unaware of the potential for FGAN detonation, did not take recommended incident response actions at the fire scene, and did not have appropriate training in hazardous materials response.

CSB found several shortcomings in federal and state regulations and standards that could reduce the risk of another incident of this type. These include the Occupational Safety and Health Administration's Explosives and Blasting Agents and Process Safety Management standards, the Environmental Protection Agency's Risk Management Program and Emergency Planning and Community Right-to-Know Act, and training provided or certified by the Texas Commission on Fire Protection and the State Firefighters' and Fire Marshals' Association of Texas

Incident Management

Why do we need Incident Management?

- Identify needed information.
- Analyze and assess the information.
- Determine strategic goals.
- Determine tactical objectives and methods.
- Identify resources that are needed.
- Develop and implement a Plan of Action.
- Evaluate and review the incident's progress

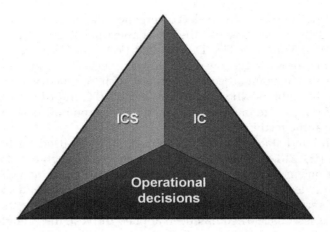

Figure 2.71 Incident Command is responsible for the management and mitigation of an incident.

Incident Command is responsible for the management and mitigation of an incident (Figure 2.71). Incident Management is the tool in the Hazmat Tool Box that the Incident Commander uses to accomplish that task. The Incident Command System is used by the IC to make operational decisions to bring the incident to a successful conclusion while keeping response personal safe.

When Ludwig Benner worked at the National Transportation Safety Board (NTSB), they did an accident risk study involving hazardous materials and concluded that emergency responders had a 10,000 times greater chance of death or injury than anyone else. This is one of the things that prompted him to develop decision models to help responders make better decisions when responding to hazardous materials.

> *Authors Note: When I spoke to him on the phone he said "the greatest challenge we face today in hazardous materials emergency response is to inform a new generation with the knowledge to safely work with those materials." One of the main reasons that prompted me to write* Hazmatology: The Science of Hazardous Materials *was much the same motivation as Ludwig Benner almost 50 years ago.*

Successful incident outcomes are directly related to the effectiveness of systems for incident management. Much of incident management involves planning and training. Incident Command organizes and facilitates the personnel responding to a hazardous materials incident to safely, effectively, and methodically function at an incident scene. It organizes the personnel present. Incident Management provides the tools for the Hazmat Tool Box to safely and effectively manage the incident.

Decision-Making Process

Data

When making decisions about what to do at a hazardous materials incident, there are three types of data that is necessary to collect before decisions can be made.

Physical

Physical information is that gathered at the scene of an incident. Your senses are utilized to search for clues and other signs of what happened, what is happening now, and what may happen in the future. Physical information is necessary before the objectives for the incident can be formulated and other decisions made.

Technical

Technical information comes from a number of sources. CHEMTREC, National Response Center, Computer Based programs like CAMEO, WISER, and others. MSDS sheets and shipping papers also provide technical information. Hard copy reference materials are also a source. Technical information is factual and pre-verified information that you can count on being accurate.

Cognitive

Cognitive information is what is in your on board data base (the one in your head). Experience, shared knowledge from others, training, and knowledge of previous and historical incidents have all contributed to your cognitive knowledge.

Recognition Primed Decision Model

Your brain is much like a computer hard drive; it has the capacity of maybe a gazillion bites of information. When you get involved in any type of incident, your brain scans it and stores the information in your on-board computer. It looks for a match of an incident you have been involved with previously or may have heard about or studied. If the hard drive finds a match, it automatically directs the behavior based on past behavior that ended with a satisfactory result. Simply, the decision maker has an idea of how things work based on the knowledge that has been gained from experience and study of past incidents. The options are compared against what is known to work.

Historical incidents, like those in volume one can be substituted for incidents you actually experience yourself. Reading and learning about them will put you there and you can experience what went wrong, if anything, then by looking at lessons learned, you can also learn what is known to produce a satisfactory result.

> **Hazmatology Point:** *March 4, 1996, eighteen years after the Waverly TN incident, a similar derailment occurred in Weyuwega, WI and Assistant Chief Jim Baehnman told me when I visited there that "he used his knowledge of the Waverly incident to formulate tactics in Weyuwega". This may have had a direct impact on the Weyuwega incident in terms of safety to emergency personnel and residents. There was not a single serious injury or death as a direct result of the derailment in Weyuwega. Assistant Chief Jim Baehnman said "from the start of the incident, the tone of the incident would be driven by safety and not time" (Figure 2.72).*

By focusing on the likely rather than the unlikely, you will be able to better handle the incidents you may be faced with and those skills will also prepare you for the unlikely events. The exception I mentioned above involves locations where you have "exotic chemicals" in a fixed facility or transported through your jurisdiction (like Houston). Even though you may have exposure to the exotic chemicals, that does not mean you will have an incident involving them. By knowing they are in your community, you can learn about them and you can prepare for the unlikely

Figure 2.72 Chief Jim Baehnman told me when I visited there that "he used his knowledge of the Waverly incident to formulate tactics in Weyuwega". (Courtesy Weyuwega Fire Department.)

which will make your response more effective if it does occur. Generally, "exotic" materials are few and far between and incidents are rare.

Traditional Decision-Making Models

Before the decision making process can proceed and models selected, all of the data gathering process must be complete and a mechanism for updated information must be in place. There are several decision making models available or you can make up your own, it doesn't matter as long as you go through a modeling process. Three have been selected for this volume. DECIDE and GEBMO were created by Ludwig Benner (1970s) and GEDAPER David Lesak (1980s). Both of these systems are still valid today.

D.E.C.I.D.E.

Ludwig Benner "The Father of Modern Hazardous Materials Thinking."

A Decisive and Innovative Change To help firefighters think through a HAZMAT situation, Benner developed an innovative decision-making process, appropriately named DECIDE.

Detect HM Presence
Determining the presence of hazardous materials is the single most important thing an emergency responder can do. It is invaluably important that emergency responders at all levels receive Operations Level training and learn how to determine the presence of hazardous materials.

Estimate Likely Harm without Intervention
This step requires gathering of information about all of the actors in the incident and analyzing the information to determine what is the likely outcome without intervention. Often times this is not a step considered by modern day emergency responders. Most feel they have to intervene because it is part of their job. Emergency responders have been injured and lost their lives over the years when intervention should not have been an option under the circumstances. Without a hazard/risk analysis and looking at harm without intervention responders may be unduly placed in harm's way just because we can.

Choose Response Objectives
Objectives are broad in nature, they need to identify what the overall incident outcome is that we are looking for. Once intervention is deemed

necessary, then objectives need to be determined indicating what our overall incident outcomes will be. Objectives do not provide specific information on how the objectives will be accomplished.

Identify Action Options
We need to determine exactly with it is we want to accomplish and determine if we have the personnel and resources to accomplish our options. Determine what options are available to be used based upon available personnel and resources. For example, if we have a flammable liquid fire and an option would be to put the fire out with foam. It would be necessary to have the amount of foam necessary to put out the volume of fire we are facing. If we do not, then it is not an option.

Do Best Option
Once all of the available options are considered, a decision needs to be made about what option is the best in order to reach the overall incident objectives. That option or options then need to be implemented.

Evaluate Progress
Determine if the options taken are making progress towards ultimate mitigation of the incident. If not, then it may be necessary to change options and try another approach and evaluate that. When options are selected without complete information, they may be doomed to fail. Evaluation and implementations of options continues on until the overall incident objective is accomplished and the incident is stabilized.

The DECIDE acronym represents key decision making points that occur during a typical HAZMAT emergency. "The intent of the DECIDE process," according to Benner, "is to help the responder get 'ahead of the curve' during a HAZMAT incident." "The goal," he emphasizes, "is to constantly update the predictions of what's going to happen next, in order to see how the actions are changing the outcome." With a HAZMAT incident, you have to focus on the outcome. The beauty of the DECIDE process is this: If you can't make a prediction about what will happen next, you can pinpoint the data gaps that will ultimately allow you to make a prediction."

Now many years removed from his days of teaching hazmat, but still interested in the health and well being of firefighters.

"Back then," Benner pointed out, "firefighters received HAZMAT training pretty much the same way, following the prevalent fire-service paradigm at the time: attack and extinguish. I wanted to change that paradigm by teaching firefighters the *importance* of thinking their way through an incident rather than jumping into the middle of something they didn't really understand. I wanted to show them how to look at a situation, interpret the visual cues, and predict what was going to happen next."

"Additionally," he continued, "my training program illustrated how critical it is to start out with a game plan, even if it's pretty basic. If the situation isn't going to create a problem, maybe you don't have to do anything. On the other hand, if it's going to hurt somebody, you have to figure out *how* it's going to hurt them and decide whether or not you can do anything about that."

Benner offers this perspective: "I had the very distinct advantage of hindsight when I was investigating accidents for the NTSB, and once you start to understand why people are doing things, you start to see what's going wrong. Back then, firefighters were using the same paradigm for HAZMAT incidents as they were for structural firefighting – and it wasn't working. All I did was show them how to look at the situation a little differently (by using the DECIDE model) and appreciate the differences between a firefighting mindset and a HAZMAT mindset. Hazmat incidents can't be handled with a cookbook approach, and I'm not a believer of teaching cookbook-type HAZMAT training – you have to use your head."

GEMBO

GEMBO is short for <u>GE</u>NEAL <u>HA</u>ZARDOUS <u>M</u>ATERIAL <u>BE</u>HAVIOR <u>MO</u>DEL This model allows an orderly assessment of events that are likely to take place when a container of hazardous materials is stressed (Figure 2.73). Hazardous materials containers in general are designed to hold the hazardous material under normal conditions of shipping, storage, and use. When containers are stressed beyond their normal capability to hold the hazardous material the container is likely to fail. When containers fail certain predictable events will take place both with the container and the hazardous material. Events may have an unfavorable impact on the public and emergency responders. Impacts may be prevented by removing the public and emergency responders out of harm's way and letting the event take its course. Ludwig Benner Jr. told me this is a time to "Go set on a hill and watch it happen, you will never see another like it." There is nothing wrong with that choice of action if it is the only safe option. However, using the GEMBO model, you can scientifically determine what actions should be taken based upon the incident circumstances when you arrive.

Anywhere on the model where it mentions Exposure, it is referring to emergency responders primarily, but that also extends to the public as well. What you want to do is change the potential outcome, which is what is likely to happen without intervention. The entire purpose of this model is to save lives of responders and the public. Property can be replaced and the environment can be cleaned up, lives cannot be replaced. This model allows for you to determine appropriate intervention points during the incident, also dependent on what is happening when you arrive. History

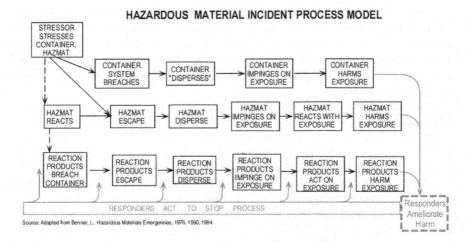

Figure 2.73 This model allows an orderly assessment of events that are likely to take place when a container of hazardous materials is stressed. (Courtesy Ludwig Benner).

has shown that effects of stressors on containers do not always cause container failure before you arrive. GEMBO provides two courses to take based upon container stressors, Container Breach or Hazmat Reactions which may lead to container breach. When you arrive either the container is breached or it is not. If not, you may be able to intervene at certain points, again based upon when you arrive.

Some of Ludwig Benner's conclusions from incident investigations for the NTSB that led to the development of his models are listed below:

- Traditional attack and extinguishment approaches did not work for hazmat emergencies.
- Firefighters were "programmed" to get into trouble at hazmat emergencies.
- Cookbook approaches to hazmat emergencies produce bad outcomes.
- There must be a better way of responding to hazmat emergencies.

According to Benner, there are two types of responses, Adaptive and Habitual (Cookbook). GEBMO allows for an adaptive approach to an incident based upon risk analysis, previous events and appropriate intervention points. Historically the vast majority of firefighter deaths and injuries at hazmat emergencies happened before the advent of decision models. That is not to say we have not lost lives since, however, we need to keep educating hazmat responders so that we do not fall into complacency.

Dealing with hazmat emergencies can be conducted safely and effectively if we are able to adapt to the situations we encounter and use models.

GEDAPER

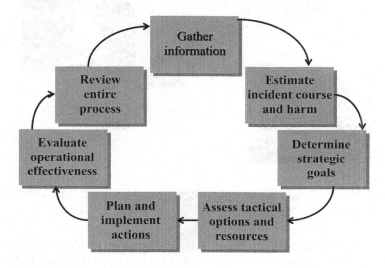

Gather Information
Accurate information is critical in determining actions to be taken at a hazardous materials incident. Information to be gathered in the decision making process includes the hazardous material or materials released, the container they are or were in, and the environment. It is necessary to identify the hazardous material(s) before mitigation tactics can be determined. Identification of the product tells the IC what the hazard or hazards are to response personnel and the public. At the very least, the hazard class of the material must be identified.

Estimate Incident Course and Harm
Once enough information has been obtained, then the IC must determine what the course, that is what is the hazardous material likely to do and harm, which determines vulnerability to responders, the public, and the environment. Then a thorough risk benefit analysis needs to be conducted to determine what actions should be taken (Figure 2.74).

Determine Strategic Goals
Based upon information gathered and potential course and harm determined, the IC must set up strategic goals to accomplish to safely mitigate the incident. Strategic goals are very broad in nature. An example might

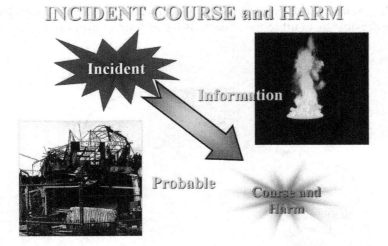

Figure 2.74 A thorough risk benefit analysis needs to be conducted to determine what actions should be taken.

be to prevent a hazardous material from reaching a vulnerable population. A number of options may available to accomplish this goal, but those are determined in the next step of the process.

Assess Tactical Options and Resources
To facilitate the implementation of Strategic Goals, each option available needs to be evaluated to determine the best option given the circumstances of the incident and the resources required to accomplish them. Options are not viable if the resources are not available to carry them out.

Plan and Implement Actions
The plan and implementation of actions will help convert the ICs ideas into actions. It identifies control methodology and personnel responsibility. The plan also assists with plan communication, transfer of command, and incident documentation. It will be specific, effective, efficient, and flexible. Plan of Action considerations must take into account existing response plans. These include the LEPC Plan, and the responding agencies Emergency Response Plan.

Evaluate Operational Effectiveness
Determine the effectiveness of the Plan of Action. Is the operation going as planned? Does it conform to standard operating procedures and guidance, standards, and safety requirements? Evaluation also determines possible errors and their severity.

Review Entire Process

Periodically review the Plan of Action, particularly during long incidents. Determine if the plan is working ok? If it is, no changes are necessary. If the plan is not working as intended, it can be modified or scrapped entirely and a new plan developed. Both options will then be run through the entire GEDAPER process again.

GEDAPER is very similar model to DECIDE but illustrates the cyclic decision-making process. Basically information that is fed into the decision model is the same. GEBMO is a hazardous materials behavior model to figure out potential course and harm in order to determine objectives. Responders need to work with each or one of their own to determine which works best for them. What is important is that a decision model is used. None of these models will be of much help at an incident if responders are not familiar with them and have used them in table top or full-scale exercises. Another way to become familiar with the models is to apply them to historical incidents. You can see if the incident outcome may have been different if models had been available and used when these incidents occurred.

These are examples of current models that are available that have been developed by others. No one model has to be followed. You can make up a model of your own that better meets the needs of your jurisdiction. The important thing is that you use some kind of decision making model to deal with a hazardous materials release successfully.

Incident Priorities

Many things are going on at one time and when you arrive on the scene of a hazardous materials release you need to set priorities. The following is a hierarchical listing of the major priorities for incident response.

- Life Safety.
- Incident Stabilization.
- Protection of Property and Environment.

Life safety is always number one and includes both responder safety and public safety. Responder safety is always the first consideration. If anything happens to responders that prevents them from carrying out public safety than no one may be able to be saved. Once life safety is assured then incident stabilization becomes the next task. There a number of ways to accomplish that stabilization, which may include withdrawing to a safe distance and letting the incident take its course. This is done when responder safety cannot be provided. One thing to remember is the hazmat teams responsibility is generally stabilization, not clean-up. Those responsible for the release are also responsible for the clean-up.

Incident Levels

Not all hazardous materials incidents are the same in terms of details and scope. Often times, jurisdictions determine levels of response that take into account what can be handled internally and what requires additional resources from other jurisdictions.

Level I is often just a response of an engine or truck company with monitoring capability. It may include clean-up of fuel spills at vehicle accidents. Certain size spills may require response of the hazmat team for additional resources.

Level II incidents usually initiate the hazmat team response based on local protocols and resources. The hazmat team may be from the local jurisdiction or mutual aid from another jurisdiction if your jurisdiction does not have one.

Level III these are what I like to call once in a lifetime or career events. They are beyond the capability of most jurisdictions to handle on their own. Multiple agencies will be needed including state and federal resources. Implementation of NIMS will also be necessary.

Scene Control Features

Establish Perimeters

Establishing the perimeters and zones is the first step to denial of entry and protecting responders and the public, and in the case of a potential criminal or terrorist scene, preserving evidence (Figure 2.75). Personnel

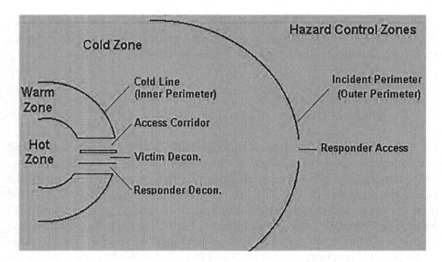

Figure 2.75 Establishing the perimeters and zones is the first step to denial of entry and protecting responders and the public.

must be assigned to prevent entry. These personnel may be law enforcement, fire police, or first responders. This assures that only appropriately trained and protected responders may enter.

Perimeter Distances
Inner perimeter established, Minimum of 150 ft.
    Upwind and further away from the incident than:

    Edge of any debris field.
    Furthest victim.
    Visible product (including vapors).

Initially, the DOT ERG is the best source of Isolation and evacuation distances. Local preplans, emergency response plans, and SOPs or SOGs. Outer perimeter should be far enough away to allow effective operations and protect the public.

Isolation Zones (Figure 2.76)

Hot Zone: Where hazardous materials are located. Because of terrain and building locations the size and shape of the Hot Zone will depend on the circumstances of the incident and the hazard of the material involved (Figure 2.77).
    Warm Zone (Contamination Reduction Zone) – Where decontamination takes place.
    Cold Zone – Everywhere else.

Isolation Zones

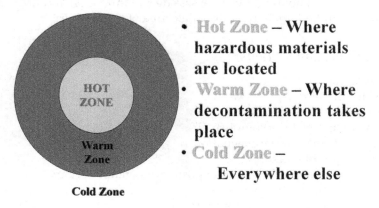

- Hot Zone – **Where hazardous materials are located**
- Warm Zone – **Where decontamination takes place**
- Cold Zone – **Everywhere else**

Figure 2.76 Isolation Zones.

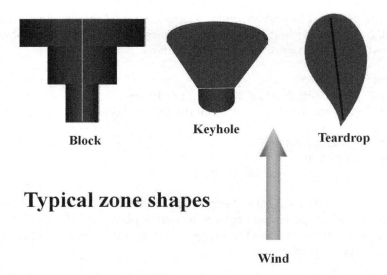

Typical zone shapes

Figure 2.77 Because of terrain and building locations the size and shape of the hot zone will depend on the circumstances of the incident.

Access to Zones

Access to zones is from upwind, upstream (and uphill if possible).

If approach and setup must be downhill or fully upwind, expand the perimeters and zones (Figure 2.78).

Safe Refuge Ares

A safe refuge must be set up at the edge of hot and warm zone.

It serves as a holding area to assess decontamination and EMS needs for victims and others in the immediate area of the release.

Public Protection Options

Evacuation.
Shelter In Place
Evacuation Considerations
Can evacuation be accomplished?
Who will perform the evacuation?
How long will it take?
Is there a potential for public exposure or contamination?
How will the public be notified?
How will the public be informed?
Sheltering Considerations

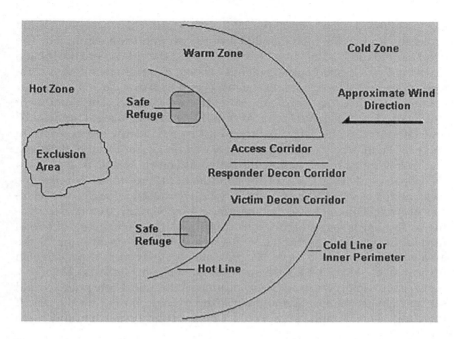

Figure 2.78 Access to zones is from upwind, upstream (and uphill if possible).

Release continue over 2 to 4 hours?
Will existing shelter provide adequate protection?
Is there potential for mass fire or explosion?
Must evacuees pass through contaminated environment?

Case Study: Crete, NE February 19, 1969, Derailment and Anhydrous Ammonia Release Shelter in Place Effectiveness Substantiated

February 18, 1969, set up as a "perfect storm" for a train derailment in Crete, Nebraska, a small college town of approximately 4,500 people.

Temperature was 4°F, wind was calm, relative humidity was 90%, approximately 14 inches of snow on the ground a temperature inversion was in place and ground fog. At 6:30 a.m. C.S.T. Chicago, Burlington, and Quincy (CB&Q) Train #64, consisting of three locomotive units and 95 cars, were entering town at 52 mph on the single main line track. Eleven box cars were standing on a siding South of the main line. Train #824 with one locomotive and 49 cars was standing on a siding North of the main line. It contained three tank cars of anhydrous ammonia. As Train #64 passed the turnout leading to the Old Wymore main siding, the spread closure allowed the wheels of the 28th car to derail. The wheels struck and broke the guard rail, then derailed and the car and train continued. The 72nd car derailed at the "frog" (switch) toward the side where the broken guard rail was located and a total of 19 cars of this train derailed. A collision occurred between cars on trains #64 and #824 and caused a tank car containing anhydrous ammonia to release its contents. The derailment site blocked completely Unona Avenue between Nebraska Highway 33 and 13th Streets. This left 13th Street East bound as only direct response route to the injured people and the only evacuation route out of the area.

Following an investigation, the National Transportation Safety Board (NTSB) attributed the cause of the derailment to "movement of rail at the turnout due to lateral forces of the locomotive caused by surface deficiencies of track. The track was not maintained for 50 mph operation according to standards and irregularities contributed to the increase of lateral forces." In other words, the Train #64 was speeding! Speed limit for the area conditions was 35 mph. Six previous derailments had occurred in or near Crete since 1954. One such derailment on Memorial Day 1963, occurred in almost the exact same location as this one. Crete police would set up radar at the location and clock the trains coming into the city. If traveling over 35 mph. they would call the train dispatcher and warn them.

Tank car SOU263210, an ammonia car split into two pieces, releasing 29,200 gallons of ammonia almost instantly, the top 16 feet landing 200 feet over Highway 33 and landed in the front yard of a residence at 1109 Highway 33 (Figure 2.79). The rest of the tank with the sill intact was propelled 140 feet onto Unona Ave.

GATX 18120 (DOT112A), shattered completely, releasing 29,200 gallons of liquid ammonia, which almost immediately turned into ammonia gas (Figure 2.80). One gallon of ammonia liquid produces 877 gallons of gas volume. Portions of the shattered tank traveled to the yards of residences located North of the derailment at (1)813, (2)905, (3)907, 13th Street and one south of Highway 33. The main group of people that died were located North and East of the pieces of tank car 18120. NTSB reported the shattering was caused by a heavy blow delivered to the head of the tank car by the coupler of another car and the brittleness of the metal at a very cold temperature, 4°F.

Casualties

It appears the highest and ultimately lethal concentrations of ammonia were located on the West end of 13th Street on the South side of the street. All of the victims were in that area when the impact occurred. Three Crete residents died during the accident, three died later in the hospital. Three unidentified transients riding the train were killed by trauma during

Figure 2.79 The 72nd car derailed at the "frog" (switch) towards the side where the broken guard rail was located and a total of 19 cars of this train derailed. (Courtesy Crete Fire Department.)

the derailment. Injury reports varied, however, the Crete News, the local paper, lists approximately 25, although the NTSB reported 53 in its final report. Of the injured were two train crew members, one, the train conductor, fell approximately 18 feet as he stepped from the train, going over a bridge West of the derailment. He was later transferred to Lincoln for treatment.

Hatchetts 725 W. 13th Street

"Ron Hatchett a 21 year old student at Doane College and star football player, his wife, Ethelene and 4 year old daughter Gloria apparently ran out of their house following the derailment. Mrs. Hatchett was found face down, unconscious in the driveway of the home. Her daughter was near her. Ron Hatchett had made it across the street to the Gottlob Rauscher home, 800 W. 13th Street. All of the Hatchett family was alive when they arrived at the Crete Hospital, however Ron and Gloria died a short time later. Mrs. Hatchett survived, possibly because she was face down when found, which might of limited the amount of inhaled ammonia she received."

Figure 2.80 The rest of the tank with the sill intact was propelled 140 feet onto Unona Ave. (Courtesy Crete Fire Department).

Erdmans between 1005 and 1045 W. 13th Street

"Louis Erdman and his wife Maxine operated the Crete Cleaners on West 13th Street Northwest of the derailment. Mrs. Erdman was talking on the phone to her daughter, Mrs. Don Wolverton at the time of the impact. Their home had been punctured by debris and they fled their home.

Mr. Erdman made it between 25 or 30 feet before falling to the place where he was found. Mrs. Erdman made it to a neighbor's home, and was later found dead."

Hoesche 1005 W. 13th Street

"Mr. Hoesche heard the crash and encountered the gas smell when he looked out the door of his house. He rushed into bedrooms to get his four children and covered their faces with wet towels. He then heard a woman screaming in the street, very likely Mrs. Erdman, and he ran out to help her into the house. Later Mrs. Erdman was found dead in the Hoesche home."

Safranek 905 W. 13th Street

Crete News reported, "The Lyle Safranek home, just to the rear and at the side of the Kovar home, was the closest to the impact." Firemen found Mr. Safranek in the street at the intersection with 13th street. He was dead.

Mrs. Safranek and one-year-old son were found a short distance from the front of the house in a snow bank. Rescue worker Clarence Busboom said "we had come past this place maybe 15 minutes earlier, but because of the thick gas hadn't been able to see anything." Mrs. Safranek and the son were hospitalized. The son was transferred to Lincoln in very serious condition. Both survived."

Svarc 813 W. 13th Street

Crete News reported that "Frank Svarc and his wife, "grandparents of Judy Svarc, and father of Firefighter Svarc," heard the crash, looked out and saw the thick gas spreading toward them. He opened the porch door and got a bucket of water, got cloths and went to the basement to await rescue. They kept dipping their hands in the water and wiping their faces and fanning the air with the cloths as well as holding them over their faces. They were rescued in about an hour by Crete firefighters and said it had seemed like an awfully long time."

Kovar 907 W. 13th Street

Information on the Kovar and Svare families was reported directly to the author by Roberta Kovar Strain and Judy Svare. "Mrs. Kovar, Roberta's mother and brother were away delivering newspapers when the derailment occurred, which may have saved their lives. They tried to get back to check on their family members, but were unable to. Parts of train cars crashed into the Kovar home puncturing the shell of the house and a window. Ammonia came into the dwelling through the openings. Mr. Kovar went out onto the porch to investigate what had happened and collapsed outside. He was taken to the hospital by rescuers, where he later died. Roberta Kovar was 19 at the time of the derailment and heard the noises from the crash. She was in her 2nd floor bedroom and covered herself with bed covers and was rescued about an hour later." The combination of covers and the fact ammonia is heavier than air likely kept the ammonia away from her and saved her life.

Svarc 915 Redwood Street

"Judy's father Firefighter Leonard Svarc lived at 915 Redwood Street, Northeast of the derailment site 3–4 blocks. He heard the fire whistle and responded to the fire house. When he found out what had happened, he tried to get back to his family to make sure they were ok, but he could not get there. Meanwhile, according his daughter Judy, "his wife Delma, sister Carol, brother Dale, and cousin Ron were still in the home. Delma got rags and wetted them and had the children put them over their faces. They went out to get in the family car to try and escape. There was a strong smell of ammonia and it burned their eyes. Once in the car, they tried to drive away, but got stuck. By that time the father had returned and took them to the Armory. Judy believes that the wet towels saved their lives." Judy and "her mother, were hospitalized for two days, her mother burned by the ammonia and Judy experiencing recurring bloody noses."

Crete Fire Department Responds

Information on the Crete Fire Department response was provided to the author directly from firefighters Everett Weilange, Chuck Henning, Arnold Henning, Loren Henning, and Chuck Vyhniek, who all responded to the incident in 1969. First reports coming into Crete emergency responders indicated a propane tank had exploded. Firefighters and Fire Chief Don Crete firefighters Loren Henning and Chuck Vyhniek were on the first fire apparatus responding to the call, which was just a few blocks West of the fire station. The 1946 American LaFrance open cab pumper they responded in is still in the department inventory and used as a public
relations apparatus. Fog and anhydrous ammonia vapors made it difficult to see what was going on.

At the time of the incident Crete fire apparatus did not have radio communications in their apparatus, so information initially had to be passed on person to person. With the very limited visibility, it was difficult to get a grasp of the scope of the incident and coordinate the
response. As the incident progressed, radios from the Sheriff's office were pressed into duty, which improved communications. Turnout gear was limited as were SCBA. Henning and Vyhniek indicated the mixture did not move and just hung low in the air. They did not know what the fog was, but some firefighters indicated there was a smell of ammonia in the air. (Anhydrous ammonia is heavier than air and tends to pool in low places and on the ground.)

Henning and Vyhniek approached the site from the East. When they arrived on scene they smelled something, but didn't know what it was so they backed out and approached from a dirt road along the tracks from the North. They smelled something there as well. They drove by Douglas Manufacturing and stopped just West. Henning jumped from the apparatus at 17th & Main Streets and started evacuating people from the area. Henning then went door to door with SCBA and several other firefighters on 13th Street. The south and west sides of town were evacuated by firefighters.

Firefighters encountered bodies and injured as they made their search. Exposed skin on victims exhibited deterioration, likely from contact with ammonia or vapors. Firefighter Arnold Henning reported driving his apparatus into the cloud and finding body parts that were later determined to be transients. Firefighter Chuck Henning worked for Wanek's Furniture Store in Crete and used a furniture delivery truck to take bodies to the mortuary. While searching, firefighters located a car that had stalled because of lack of oxygen, which had been displaced by the ammonia gas. Mr. Alvin Rozdalousky got out of the car and ran 5 blocks East to Main Street and safety. Firefighter Everet Weilage used his own vehicle to alert people in the South side of Crete and move many from harm's way. Also helped out at the station. All of the firefighters I talked with indicated their first priority was to evacuate people to safety, and isolate the scene. Between 200 and 300 people were evacuated from the area. Crete firefighters had previously trained and participated in table top exercises for disasters. This preparation likely helped them as they responded to the train derailment.

Following evacuation, some people went to be with relatives; others to Doane College in Crete; the Armory; and others to the fire station. Food was brought in by residents and businesses and supported by the Red Cross. Nebraska National Guard from Lincoln assisted in a secondary search to determine if all had been rescued and evacuated. Later in the morning, a helicopter was brought in to and helped to disperse the gas fumes with its whirling blades. In spite of the conditions at the scene and lack of protective equipment, only four firefighters were injured in the incident from ammonia exposure and were treated and released.

Firefighters remained on scene for three days. According to Wally Barnett, assistant state fire marshal, "If there had been a West wind and it had been a clear day that cloud of gas could have made a clean sweep of the town." Assistance from other towns, fire, and police came from Lincoln, Malcolm, York, Seward, Milford, Southeast Rural District in Lincoln, Hallam, Friend, Beatrice, Fairbury and Wilbur. Fourteen Crete firefighters who responded to the incident are still living at the time of this interview.

Crete's encounter with the train derailment and anhydrous ammonia occurred prior to the development of organized response to hazardous materials in the U.S. Fire Service. United States Department of Transportation (DOT) did not have an Emergency Response Guide Book or a placard and labeling system in place. No markings of the dangerous chemicals in transportation were required on the tanks of anhydrous ammonia. Requirements did not come until the early 1970s. DOT had also not yet coined today's term for chemicals in transportation, Dangerous Goods or Hazardous Materials as they are normally called.

The idea of sheltering people in place inside of buildings against chemical exposure did not exist at this time. However, many people did in fact shelter themselves inside their homes, placed wet rags over their faces and some covered with blankets. Not only did they shelter in place and protect themselves from the ammonia, but saved their lives because of their inadvertent actions. Not a single person who stayed inside the entire time of the emergency and took self-protective actions, died or was

Figure 2.81 GATX 18120 (DOT112A), shattered completely, releasing 29,200 gallons of liquid ammonia, which almost immediately turned into ammonia gas. (Courtesy Crete Fire Department.)

seriously injured. Quite a bit of common sense was involved in the thinking of members of the public, which helped keep them safe.

Those who died from ammonia exposure, died when they left their homes and were overcome by the ammonia vapors outside (Figure 2.81). Some died on their driveways and one on a street corner, beyond the apparent safety of their homes. Curiosity, called them out to see what had happened and they paid the ultimate price. One person who was inside their residence and quickly went onto their porch, turned around and went back inside. They still had enough exposure to the ammonia that they died of complications later in the hospital. Unknowingly, even before the concept of Shelter in Place had been developed, victims of the Crete derailment confirmed the effectiveness of Sheltering in Place when hazardous materials are released outside of buildings or a vapor cloud travels to populated areas preventing an expeditious and safe evacuation.

It is interesting to note that the house number 1005 13th Avenue is located adjacent to the cleaners where two victims died and just West of 13th and Unona where one person died. The person at 13th and Unona who died on the street corner, lived at 813 13th Street. The house number "1005," is the DOT designation for anhydrous ammonia in bulk quantities, which was not used at the time of the derailment.

Had this incident happened at any other time of the year, it would have likely been much worse. In warmer weather, people would have had less clothing on, windows might have been opened in the dwellings more people may have gone outside to see what happened. Liquid ammonia has a boiling point of −40°F below zero. Anhydrous means without water. Ammonia seeks water when released into the environment. However, at 4°F, there wasn't much water in the area of the derailment that was not frozen. It is unlikely that people exposed to the ammonia were sweating if they had been ammonia could have reacted with the moisture on the skin causing serious burns. Even though the air temperature and items in the environment were above the boiling point of ammonia, warmer temperatures would have caused even more gas to be formed quicker by the spilled liquid ammonia. Not everyone who went outside died, however no one who sheltered in place died. The snow on the ground may have someway created a barrier to the exposed skin and airways of those who survived.

During the following week after the derailment, the Crete Fire Department was dispatched to a fire at Doane College Merrill Hall for a fire. Firefighters laid hose lines and began to fight the fire, when the hoses failed and started to leak from exposure to ammonia at the derailment. One firefighter commented the hose looked like a soaker hose with all the leaks. Crete does not have a hazardous materials response team. All firefighters are trained to the hazmat operations level. If a hazmat incident occurs in Crete or the surrounding area teams are available from Lincoln and Beatrice (*Firehouse Magazine*).

Crisis Management

Response organizations deal with an incident or incidents of an unexpected scope that threatens responders and the public. It is generally not a routine incident that the responders deal with on a regular basis.

Consequence Management

Efforts made to protect public health and safety, restore essential services, and provide emergency relief to public and private organizations and members of the general public affected by the consequences of a fire, hazardous materials release, or terrorist attack.

Emergency Operations Center (EOC)

The Emergency Operations Center is a building that houses the Emergency Manager and staff for a city, county, state or in the case of the Federal Government, FEMA has an emergency operations center (Figure 2.82). The purpose of the EOC is to provide support for large-scale incidents that may happen within a community. These include floods, tornados, hurricanes, and others. EOC also supports hazardous materials incidents when

Figure 2.82 The Emergency Operations Center is a building that houses the Emergency Manager and staff for a city, county, state.

requested to do so my an Incident Commander. The EOC is a resource for the Incident Commander. State EOCs have compacts with other states and can request and receive resources from other states quickly during an emergency without going through a state's governor. EOCs can also seek Federal Disaster Assistance from FEMA by requesting a declaration of a disaster from the President of the United States. This can trigger Federal resources and funding for local incidents.

When an EOC is activated during an emergency, representatives from local government, including community leaders and from state government including the governor are represented at the EOC to enhance quick decision making and allocating and deployment of needed resources. The EOC houses and implements community disaster plans when needed. EOC generally handles requests and implements evacuations and shelter in place orders issued by Incident Command. They can also notify the community at large by a number of resources including media, outdoor sirens, door-to-door notifications, and others.

Activation of the EOC early on in a major incident can ensure the availability of needed resources as an incident unfolds.

Incident Generated Plans

Incident generated plans are based upon the local emergency plans including the LEPC, facility plans, agency response plans, and others. The better the pre-incident planning process, the easier the incident generated plans are to complete.

Site Safety Plan

Documents Plan of Action chosen and SOP's used.
Tracks activities.
Assures safe performance by appropriate staff.
Sets up checklists (Figure 2.83).
Is divided by ICS functions.
Assures compilation and documentation of data.

Plan of Action

Identifies
    Control methodology.
    Personnel responsibility.
Assists With
    Plan communication.
    Transfer of command.
Incident documentation.

Site Safety Plan Check Lists

Figure 2.83 Site Safety Plan Check Lists.

Plan of Action Considers Existing Response Plans

LEPC Plan
  Basic plan assigns agency responsibilities.
  Facility plans have information on planning facilities.
Employer's ERP

  Identifies as specific agencies and response teams will operate.
  Generally include operational procedures (similar to SOG's).
  If done well, provides responsibilities of all personnel roles and
    responsibilities.
  Can act as training guidelines.

Plan of Action

Specific.
  Effective.
  Efficient.
  Flexible.
  Helps convert the IC's ideas into actions.

Terminating the Incident

Recovery

    Returning as closely as possible to pre-incident conditions.

Termination

    Process of documenting, Identifying, and resolving deficiencies (Figure 2.84).

Debriefing

    Performed immediately after the incident.

    Designed as fact finding and documentation.

    May include CISD.

        Identifies.

         Specific occurrences.

        Personnel involved.

        Degree of success.

        Develops an accurate timeline.

        Is developed through group process.

Critique

    A method to identify and document accomplishments, problems, and shortcomings.

    Focus on developing recommendations for system improvements and additions.

Figure 2.84 Termination of the incident is the process of documenting, Identifying, and resolving deficiencies.

Characteristics:
    Professional, not personal.
    Information oriented.
    Positive as well as negative.
    Document recommendations.

After Action Analysis

A review and analysis of the debriefing and the critique.
Helps identify:
    Specific trends.
    A need to develop training, SOP's and resources.
    Needed personnel levels.
    Individuals responsible for follow-up.

After Action Report

They are compilations of incident documentation, debriefings, critiques, and after-action analysis.
Elements:
    Personnel and medical file updates.
    Equipment exposure updates.
    Operational incident reports.
    Staff incident reports.

After Action Follow-Up

They are compilations of incident documentation, debriefings, critiques, and after-action analysis.
Elements:
    Personnel and medical file updates.
    Equipment exposure updates.
    Operational incident reports.
    Staff incident reports.

Local Emergency Planning Committee (LEPC)

LEPC Community Emergency Response Plan

On December 4, 1984, a tragic accident occurred in Bhopal, India, which killed over 3,000 people. Methyl isocyanate was released from a chemical plant and spread into populated areas of Bhopal. As a direct result of that accident, the Congress of the United States passed the Emergency Planning and Community Right-to-Know Act of 1986 (EPCRA). One of the main elements of this act is the formation of State Emergency Response Commission (SERC) and Local Emergency Planning

Committee (LEPC). The governor of each state is required to appoint an SERC to carry out the mandates of EPCRA. Each SERC, through the governor, appoints members of the LEPC and coordinates the activities to ensure consistency among LEPCs and compliance with the mandates of EPCRA.

Each LEPC is responsible for developing a plan to deal with releases of extremely hazardous substances (EHSs), as identified by the U.S. Environmental Protection Agency (EPA), within their jurisdiction. LEPC jurisdictions are determined by the SERC. The location of the LEPC jurisdiction varies from state to state. Some states set up jurisdictions consisting of several counties, while others identify individual counties or cities as planning districts. Part of the LEPC planning process includes identification of transportation routes that are used in the transportation of EHS. In fact, in some cases, the only hazardous materials posing a threat to a community are the ones being transported through it. Each LEPC plan identifies those routes and based upon the results of the commodity flow study (CFS) conducts risk assessments and vulnerability studies for the areas affected. This will allow for a more knowledgeable and safer response to incidents by emergency responders.

Local Emergency Planning Committees

Under the Emergency Planning and Community Right-to-Know Act (EPCRA), Local Emergency Planning Committees (LEPCs must develop an emergency response plan, review the plan at least annually, and provide information about chemicals in the community to citizens. Plans are developed by LEPCs with stakeholder participation. There is one LEPC for each of the more than 3,000 designated local emergency planning districts. The LEPC membership must include (at a minimum):

- Elected state and local officials
- Police, fire, civil defense, and public health professionals
- Environment, transportation, and hospital officials
- Facility representatives
- Representatives from community groups and the media

What Are the Required Elements of a Community Emergency Response Plan?

- Identification of facilities and transportation routes of extremely hazardous substances
- Description of emergency response procedures, on- and off-site

- Designation of a community coordinator and facility emergency coordinator(s) to implement the plan
- Outline of emergency notification procedures
- Description of how to determine the probable affected area and population by releases
- Description of local emergency equipment and facilities and the persons responsible for them
- Outline of evacuation plans
- A training program for emergency responders (including schedules)
- Methods and schedules for exercising emergency response plans

Commodity Flow Study

The Emergency Planning and Community Right-To-Know Act of 1986 (EPCRA) mandates Local Emergency Planning Committees (LEPC) to identify transportation routes used to transport extremely hazardous substances (EHS). Every community, large and small, has hazardous materials transported around or through it. Transportation incidents involving hazardous materials can be difficult to handle and even more difficult to plan for. One-way of reducing the difficulty is to conduct a transportation commodity flow study. Such a study identifies the types of hazardous materials that are routinely transported on a particular transportation route. This knowledge will assist planners and emergency responders in planning for and dealing with incidents involving hazardous materials.

There are various consulting firms that can conduct commodity flow studies and make reports to local planners and emergency responders. The problem is that these studies cost money. There was not any funding tied to EPCRA. There have been small amounts of money made available through the Hazardous Materials Emergency Preparedness Program (HMEP) Grant Program for commodity flow studies. However, these funds may not be adequate to fund a consultant to conduct the study. A less expensive alternative may be to use volunteer or existing agency personnel to conduct the traffic survey for the study (Figure 2.85).

Conducting a Commodity Flow Study

Highway Gathering accurate survey information is very important to conducting an effective survey that provides information useful to the LEPC in doing risk assessment (Figure 2.86). According to the U.S. Department of Transportation (DOT), almost 50% of highway transportation vehicles carry hazardous materials. Only 50% of those are required to display placards because of weight requirements. Most of these vehicles

Figure 2.85 A less expensive alternative for conducting a commodity flow study may be to use volunteer or existing agency personnel to conduct the traffic survey for the study.

are box trailers. It is important to count all box trailers even if they don't have placards. Tanker trucks carrying hazardous materials will almost always be marked or placarded.

Listed below is a checklist of tips and information to help in conducting a placard survey of highway, rail, and water transportation routes:

- Conduct the survey from a safe location.
- Count all box trailers and intermodal containers on the highway.
- Note highway vehicles with placards.
- Note trucking company names.
- Note stenciled commodity names.
- Note type of container/tank whether pressure or nonpressure.

Railroad Railcars are required to be placarded, regardless of the quantity of hazardous materials present (Figure 2.87). That requirement makes a survey of the railroad easier to do. Only identify railcars with placards.

- Identify the cars with placards.
- Note the type of car, i.e. box car, pressure tanker, and liquid tanker.
- Note any stenciling of names on tank cars.
- List any markings on railcars, such as names of manufacturers and chemical companies.

Water Transportation Hazardous materials can be found on navigable rivers, ports, and the open seas. Unfortunately, water transportation vessels are not placarded (Figure 2.88). Shipping containers may be

Figure 2.86 Gathering accurate survey information is very important to conducting an effective survey that provides information useful to the LEPC in doing riskassessment.

Figure 2.87 Railcars are required to be placarded, regardless of the quantity of hazardous materials present.

placarded, and the placards could be visible if the containers are on the deck of the ship or barge. Many times though the containers are below deck and placards are not visible. Barges that carry hazardous materials will fly a red pinnate indicating hazardous materials on board. The pinnate does not identify the hazard class or type of hazardous materials. It is possible to determine if barges are carrying liquids, solids, or gases. The best bet for obtaining information about water shipments is to visit the port facility and identify the types of containers and placards shown on those containers sitting on the dock. You may also be able to obtain the information from the shipping company.

Air Transportation Federal Express, UPS Air, and other air freight companies transport hazardous materials as permitted by the Federal Aeronautics Administration (FAA) and the United States Department of Transportation DOT (Figure 2.89). Passenger flights are also allowed to transport a further restricted group of hazardous materials. Checking regulatory requirements for air shipments would begin with both FAA and DOT regulations. Airlines may be less likely to share information on air shipments, unless an accident occurs while on the ground during the transport process, which can involve hazardous materials shipped or fuel from the aircraft. Because of the destruction of an air crash, there is not much hazard at a crash scene but the fuel in the aircraft.

Figure 2.88 Hazardous materials can be found on navigable rivers, ports, and on the open seas. Unfortunately, water transportation vessels are not placarded.

Figure 2.89 Federal Express, UPS Air, and other air freight companies transport hazardous materials as permitted by the Federal Aeronautics Administration FAA and the United States Department of Transportation (DOT).

Gathering accurate survey samples and ensuring the safety of survey personnel is important to a safe and effective survey. Make sure that the vantage point of conducting the survey is located away from traffic patterns and danger to personnel. Box trailers account for as much as half of hazardous materials shipments. Many are not placarded because they do not meet the weight requirements for placarding.

Placards are the primary method of determining hazard classes of hazardous materials in transportation. Make note of the placard color and any four-digit identification numbers on the placards. Vehicle types can add additional information to placards and labels, such as the physical state or in some cases the hazard class. Identifying the trucking company can help with obtaining additional information, especially on shipments that do not require placards. Some types of products are marked on the container with the name of the chemical or hazardous material. That is better than having a placard color or container type. Be sure and watch for stenciled names.

Make note of the types of railcars that are placarded. Most tanker cars carrying hazardous materials will have a placard with a four-digit identification number. Make note of these numbers as this can lead to the identity of the product in the container. Note commodity names stenciled on containers. Make note if the railcar is a pressure car or a liquid car. Also note other special cars earlier discussed (FEMA, OSHA, EPA, DOT).

Commodity Flow Study Highway Survey Checklist

Route _____ Direction _____ Date _____ Time-Start _____

Placard Color _____ Container Type _____

UN ID Number_____ Pressure _____ Non-Pressure _____

Trucking Company Name _____

Special Information _____
_____

Placard Color _____ Container Type _____

UN ID Number_____ Pressure _____ Non-Pressure _____

Trucking Company Name _____

Special Information _____
_____

Placard Color _____ Container Type _____

UN ID Number_____ Pressure _____ Non-Pressure _____

Trucking Company Name _____

Special Information _____
_____

Placard Color _____ Container Type _____

UN ID Number_____ Pressure _____ Non-Pressure _____

Trucking Company Name _____

Special Information _____
_____

Placard Color _____ Container Type _____

UN ID Number_____ Pressure _____ Non-Pressure _____

Trucking Company Name _____

Special Information _____

Time-Stop _____   Page _____ of _____

Commodity Flow Study Railroad Survey Checklist

Location _____ Direction _____ Date _____ Time-Start _____

Placard Color _____ Railcar Type _____

UN ID Number_____ Pressure _____ Non-Pressure _____

Special Information _____

_____

Placard Color _____ Railcar Type _____

UN ID Number_____ Pressure _____ Non-Pressure _____

Special Information _____

_____

Placard Color _____ Railcar Type _____

UN ID Number_____ Pressure _____ Non-Pressure _____

Special Information _____

_____

Placard Color _____ Railcar Type _____

UN ID Number_____ Pressure _____ Non-Pressure _____

Special Information _____

_____

Placard Color _____ Railcar Type _____

UN ID Number_____ Pressure _____ Non-Pressure _____

Special Information _____

_____

Placard Color _____ Railcar Type _____

UN ID Number_____ Pressure _____ Non-Pressure _____

Special Information _____

Time-Stop _____   Page _____ of _____

Site Specific Plans

Elements to Include in Facility Response Plan

What key elements should be included in a Facility Response Plan (FRP)? As you prepare your FRP, be sure that your plan includes the following elements:

- Emergency Response Action Plan (an easily accessible stand-alone section of the overall plan) including the identity of a qualified individual with the authority to implement removal actions
- Facility name, type, location, owner, and operator information
- Emergency notification, equipment, personnel, evidence that equipment and personnel are available (by contract or other approved means), and evacuation information
- Identification and evaluation of potential discharge hazards and previous discharges
- Identification of small, medium, and worst-case discharge scenarios and response actions
- Description of discharge detection procedures and equipment
- Detailed implementation plans for containment and disposal
- Facility and response self-inspection; training, exercises, and drills; and meeting logs
- Diagrams of facility and surrounding layout, topography, evacuation paths, and drainage flow paths
- Security measures, including fences, lighting alarms, guards, emergency cutoff valves, and locks
- Response Plan cover sheet (form with basic information concerning the facility)

Site Specific Emergency Response Plan (ERP)
OSHA 1910.120(p)(8)
Emergency response program
1910.120(p)(8)(i)

Emergency response plan. An emergency response plan shall be developed and implemented by all employers. Such plans need not duplicate any of the subjects fully addressed in the employer's contingency planning required by permits, such as those issued by the U.S. Environmental Protection Agency, provided that the contingency plan is made part of the emergency response plan. The emergency response plan shall be a written portion of the employer's safety and health program required in paragraph (p)(1) of this section. Employers who will evacuate their employees from the worksite location when an emergency occurs and who do not permit any of their employees to assist in handling the emergency are exempt from the requirements of paragraph (p)(8) if they provide an emergency action plan complying with 29 CFR 1910.38.

1910.120(p)(8)(ii)

Elements of an emergency response plan. The employer shall develop an emergency response plan for emergencies which shall address, as a minimum, the following areas to the extent that they are not addressed in any specific program required in this paragraph:

1910.120(p)(8)(ii)(A)
Pre-emergency planning and coordination with outside parties.
1910.120(p)(8)(ii)(B)
Personnel roles, lines of authority, training, and communication.
1910.120(p)(8)(ii)(C)
Emergency recognition and prevention.
1910.120(p)(8)(ii)(D)
Safe distances and places of refuge.
1910.120(p)(8)(ii)(E)
Site security and control.
1910.120(p)(8)(ii)(F)
Evacuation routes and procedures.
1910.120(p)(8)(ii)(G)
Decontamination procedures.
1910.120(p)(8)(ii)(H)
Emergency medical treatment and first aid.
1910.120(p)(8)(ii)(I)
Emergency alerting and response procedures.
1910.120(p)(8)(ii)(J)
Critique of response and follow-up.
1910.120(p)(8)(ii)(K)
PPE and emergency equipment.
1910.120(p)(8)(iii)
Training
1910.120(p)(8)(iii)(A)

Training for emergency response employees shall be completed before they are called upon to perform in real emergencies. Such training shall include the elements of the emergency response plan, standard operating procedures the employer has established for the job, the personal protective equipment to be worn, and procedures for handling emergency incidents.

Exception #1: An employer need not train all employees to the degree specified if the employer divides the work force in a manner such that a sufficient number of employees who have responsibility to control emergencies have the training specified, and all other employees, who may first respond to an emergency incident, have sufficient awareness training to recognize that an emergency response situation exists and that they are instructed in that case to summon the fully trained employees and not attempt control activities for which they are not trained.

Exception #2: An employer need not train all employees to the degree specified if arrangements have been made in advance for an outside fully trained emergency response team to respond in a reasonable period and all employees, who may come to the incident first, have sufficient awareness training to recognize that an emergency response situation exists and they have been instructed to call the designated outside fully trained emergency response team for assistance.

1910.120(p)(8)(iii)(B)

Employee members of TSD facility emergency response organizations shall be trained to a level of competence in the recognition of health and safety hazards to protect themselves and other employees. This would include training in the methods used to minimize the risk from safety and health hazards, in the safe use of control equipment, in the selection and use of appropriate personal protective equipment, in the safe operating procedures to be used at the incident scene, in the techniques of coordination with other employees to minimize risks, in the appropriate response to over exposure from health hazards or injury to themselves and other employees, and in the recognition of subsequent symptoms which may result from over exposures.

1910.120(p)(8)(iii)(C)

The employer shall certify that each covered employee has attended and successfully completed the training required in paragraph (p)(8)(iii) of this section or shall certify the employee's competency for certification of training shall be recorded and maintained by the employer.

1910.120(p)(8)(iv)

Procedures for Handling Emergency Incidents.

1910.120(p)(8)(iv)(A)

In addition to the elements for the emergency response plan required in paragraph (p)(8)(ii) of this section, the following elements shall be included for emergency response plans to the extent that they do not repeat any information already contained in the emergency response plan:

1910.120(p)(8)(iv)(A)(1)

Site topography, layout, and prevailing weather conditions.

1910.120(p)(8)(iv)(A)(2)

Procedures for reporting incidents to local, state, and federal governmental agencies.

1910.120(p)(8)(iv)(B)

The emergency response plan shall be compatible and integrated with the disaster, fire, and/or emergency response plans of local, state, and federal agencies.

1910.120(p)(8)(iv)(C)

The emergency response plan shall be rehearsed regularly as part of the overall training program for site operations.

1910.120(p)(8)(iv)(D)

The site emergency response plan shall be reviewed periodically and, as necessary, be amended to keep it current with new or changing site conditions or information.

1910.120(p)(8)(iv)(E)

An employee alarm system shall be installed in accordance with 29 CFR 1910.165 to notify employees of an emergency situation, to stop work activities if necessary, to lower background noise in order to speed communication, and to begin emergency procedures.

1910.120(p)(8)(iv)(F)

Based upon the information available at time of the emergency, the employer shall evaluate the incident and the site response capabilities and proceed with the appropriate steps to implement the site emergency response plan.

OSHA 1910.120 Emergency Response Plan

(q)

Emergency response program to hazardous substance releases. This paragraph covers employers whose employees are engaged in emergency response no matter where it occurs, except that it does not cover employees engaged in operations specified in paragraphs (a)(1)(i) through (a)(1)(iv) of this section. Those emergency response organizations that have developed and implemented programs equivalent to this paragraph for handling releases of hazardous substances pursuant to Section 303 of the Superfund Amendments and Reauthorization Act of 1986 (Emergency Planning and Community Right-to-Know Act of 1986, 42 U.S.C. 11003) shall be deemed to have met the requirements of this paragraph.

(q)(1)

An emergency response plan shall be developed and implemented to handle anticipated emergencies prior to the commencement of emergency response operations. The plan shall be in writing and available for inspection and copying by employees, their representatives, and OSHA personnel. Employers who will evacuate their employees from the danger area when an emergency occurs, and who do not permit any of their employees to assist in handling the emergency, are exempt from the requirements of this paragraph if they provide an emergency action plan in accordance with 29 CFR 1910.38.

(q)(2)

Elements of an emergency response plan. The employer shall develop an emergency response plan for emergencies which shall address, as a minimum, the following areas to the extent that they are not addressed in any specific program required in this paragraph:

(q)(2)(i)
Pre-emergency planning and coordination with outside parties.
(q)(2)(ii)
Personnel roles, lines of authority, training, and communication.
(q)(2)(iii)
Emergency recognition and prevention.
(q)(2)(iv)
Safe distances and places of refuge.
(q)(2)(v)
Site security and control.
(q)(2)(vi)
Evacuation routes and procedures.
(q)(2)(vii)
Decontamination.
(q)(2)(viii)
Emergency medical treatment and first aid.
(q)(2)(ix)
Emergency alerting and response procedures.
(q)(2)(x)
Critique of response and follow-up.
(q)(2)(xi)
PPE and emergency equipment.
(q)(2)(xii)

Emergency response organizations may use the local emergency response plan or the state emergency response plan or both, as part of their emergency response plan to avoid duplication. Those items of the emergency response plan that are being properly addressed by the SARA Title III plans may be substituted into their emergency plan or otherwise kept together for the employer and employee's use.

(q)(3)(i)

The senior emergency response official responding to an emergency shall become the individual in charge of a site-specific incident command system (ICS). All emergency responders and their communications shall be coordinated and controlled through the individual in charge of the ICS assisted by the senior official present for each employer.

(q)(3)(i)–

The "senior official" at an emergency response is the most senior official on the site who has the responsibility for controlling the operations at the site. Initially, it is the senior officer on the first-due piece of responding emergency apparatus to arrive on the incident scene. As more senior officers arrive (i.e., battalion chief, fire chief, state law enforcement official, site coordinator, etc.), the position is passed up the line of authority which has been previously established (OSHA).

> *Hazmatology Point: A friend of mine who was a chief of a large major city department before he retired told me about what he went through with OSHA and the fact that he was on scene of a major incident and did not assume command as required by this section of the OSHA regulations. The chief went through quite a bit of regulatory and court headaches along with court and attorneys. If you do not want to take command at the scene of a hazmat incident, do not show up, watch the news footage on television.*

(q)(3)(vii)

The individual in charge of the ICS shall designate a safety officer, who is knowledgeable in the operations being implemented at the emergency response site, with specific responsibility to identify and evaluate hazards and to provide direction with respect to the safety of operations for the emergency at hand.

(q)(3)(viii)

When activities are judged by the safety officer to be an IDLH and/or to involve an imminent danger condition, the safety officer shall have the authority to alter, suspend, or terminate those activities. The safety official shall immediately inform the individual in charge of the ICS of any actions needed to be taken to correct these hazards at the emergency scene.

(q)(3)(ix)

After emergency operations have terminated, the individual in charge of the ICS shall implement appropriate decontamination procedures.

(q)(3)(x)

When deemed necessary for meeting the tasks at hand, approved self-contained compressed air breathing apparatus may be used with approved cylinders from other approved self-contained compressed air breathing apparatus, provided that such cylinders are of the same capacity and pressure rating. All compressed air cylinders used with self-contained breathing apparatus shall meet U.S. Department of Transportation and National Institute for Occupational Safety and Health criteria (Figure 2.90).

(q)(4)

Skilled support personnel. Personnel, not necessarily an employer's own employees, who are skilled in the operation of certain equipment, such as mechanized earth moving or digging equipment or crane and hoisting equipment, and who are needed temporarily to perform immediate

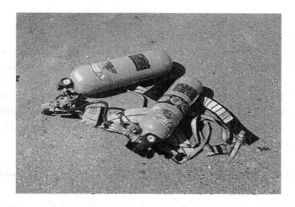

Figure 2.90 All compressed air cylinders used with self-contained breathing apparatus shall meet U.S. Department of Transportation and National Institute for Occupational Safety and Health criteria.

emergency support work that cannot reasonably be performed in a timely fashion by an employers own employees; and who will be or may be exposed to the hazards at an emergency response scene are not required to meet the training required in this paragraph for the employer's regular employees. However, these personnel shall be given an initial briefing at the site prior to their participation in any emergency response. The initial briefing shall include instruction in the wearing of appropriate personal protective equipment, what chemical hazards are involved, and what duties are to be performed. All other appropriate safety and health precautions provided to the employer's own employees shall be used to assure the safety and health of these personnel.

(q)(5)

Specialist employees. Employees who, in the course of their regular job duties, work with and are trained in the hazards of specific hazardous substances and who will be called upon to provide technical advice or assistance at a hazardous substance release incident to the individual in charge shall receive training or demonstrate competency in the area of their specialization annually.

(q)(6)

Training. Training shall be based on the duties and function to be performed by each responder of an emergency response organization. The skill and knowledge levels required for all new responders, those hired after the effective date of this standard, shall be conveyed to them through training before they are permitted to take part in actual emergency operations on an incident. Employees who participate, or are expected to participate, in emergency response shall be given training in accordance with the following paragraphs:

(q)(6)(i)

First responder awareness level. First responders at the awareness level are individuals who are likely to witness or discover a hazardous substance release and who have been trained to initiate an emergency response sequence by notifying the proper authorities of the release. They would take no further action beyond notifying the authorities of the release. First responders at the awareness level shall have sufficient training or have had sufficient experience to objectively demonstrate competency in the following areas:

- **(q)(6)(i)(A)** An understanding of what hazardous substances are and the risks associated with them in an incident.
- **(q)(6)(i)(B)** An understanding of the potential outcomes associated with an emergency created when hazardous substances are present.
- **(q)(6)(i)(C)** The ability to recognize the presence of hazardous substances in an emergency.
- **(q)(6)(i)(D)** The ability to identify the hazardous substances, if possible.
- **(q)(6)(i)(E)** An understanding of the role of the first responder awareness individual in the employer's emergency response plan including site security and control and the U.S. Department of Transportation's Emergency Response Guidebook.
- **(q)(6)(i)(F)** The ability to realize the need for additional resources, and to make appropriate notifications to the communication center.
- **(q)(6)(ii)**
- *First responder operations level.* First responders at the operations level are individuals who respond to releases or potential releases of hazardous substances as part of the initial response to the site for the purpose of protecting nearby persons, property, or the environment from the effects of the release. They are trained to respond in a defensive fashion without actually trying to stop the release. Their function is to contain the release from a safe distance, keep it from spreading, and prevent exposures. First responders at the operations level shall have received at least 8 h of training or have had sufficient experience to objectively demonstrate competency in the following areas in addition to those listed for the awareness level and the employer shall so certify:
- **(q)(6)(ii)(A)** Knowledge of the basic hazard and risk assessment techniques.
- **(q)(6)(ii)(B)** Know how to select and use proper personal protective equipment provided to the first responder operations level.
- **(q)(6)(ii)(C)** An understanding of basic hazardous materials terms.

- **(q)(6)(ii)(D)** Know how to perform basic control, containment, and/or confinement operations within the capabilities of the resources and personal protective equipment available with their unit.
- **(q)(6)(ii)(E)** Know how to implement basic decontamination procedures.
- **(q)(6)(ii)(F)** An understanding of the relevant standard operating procedures and termination procedures.
- **(q)(6)(iii)**
- *Hazardous materials technician.* Hazardous materials technicians are individuals who respond to releases or potential releases for the purpose of stopping the release. They assume a more aggressive role than a first responder at the operations level in that they will approach the point of release in order to plug, patch, or otherwise stop the release of a hazardous substance. Hazardous materials technicians shall have received at least 24 h of training equal to the first responder operations level and in addition have competency in the following areas and the employer shall so certify:
- **(q)(6)(iii)(A)** Know how to implement the employer's emergency response plan.
- **(q)(6)(iii)(B)** Know the classification, identification and verification of known and unknown materials by using field survey instruments and equipment.
- **(q)(6)(iii)(C)** Be able to function within an assigned role in the Incident Command System.
- **(q)(6)(iii)(D)** Know how to select and use proper specialized chemical personal protective equipment provided to the hazardous materials technician.
- **(q)(6)(iii)(E)** Understand hazard and risk assessment techniques.
- **(q)(6)(iii)(F)** Be able to perform advance control, containment, and/or confinement operations within the capabilities of the resources and personal protective equipment available with the unit.
- **(q)(6)(iii)(G)** Understand and implement decontamination procedures.
- **(q)(6)(iii)(H)** Understand termination procedures.
- **(q)(6)(iii)(I)** Understand basic chemical and toxicological terminology and behavior.
- **(q)(6)(iv)**
- *Hazardous materials specialist.* Hazardous materials specialists are individuals who respond with and provide support to hazardous materials technicians. Their duties parallel those of the hazardous materials technicians; however, those duties require a more directed or specific knowledge of the various substances they may be called upon to contain. The hazardous materials specialist would also act as the site liaison with Federal, state, local and other government authorities in regards to site activities. Hazardous materials

specialists shall have received at least 24 hours of training equal to the technician level and in addition have competency in the following areas and the employer shall so certify:

- **(q)(6)(iv)(A)** Know how to implement the local emergency response plan.
- **(q)(6)(iv)(B)** Understand classification, identification and verification of known and unknown materials by using advanced survey instruments and equipment.
- **(q)(6)(iv)(C)** Know the state emergency response plan.
- **(q)(6)(iv)(D)** Be able to select and use proper specialized chemical personal protective equipment provided to the hazardous materials specialist.
- **(q)(6)(iv)(E)** Understand in-depth hazard and risk techniques.
- **(q)(6)(iv)(F)** Be able to perform specialized control, containment, and/or confinement operations within the capabilities of the resources and personal protective equipment available.
- **(q)(6)(iv)(G)** Be able to determine and implement decontamination procedures.
- **(q)(6)(iv)(H)** Have the ability to develop a site safety and control plan.
- **(q)(6)(iv)(I)** Understand chemical, radiological and toxicological terminology and behavior.
- **(q)(6)(v)**
- *On-scene incident commander.* Incident commanders, who will assume control of the incident scene beyond the first responder awareness level, shall receive at least 24 h of training equal to the first responder operations level and in addition have competency in the following areas and the employer shall so certify:
- **(q)(6)(v)(A)** Know and be able to implement the employer's incident command system.
- **(q)(6)(v)(B)** Know how to implement the employer's emergency response plan.
- **(q)(6)(v)(C)** Know and understand the hazards and risks associated with employees working in chemical protective clothing.
- **(q)(6)(v)(D)** Know how to implement the local emergency response plan.
- **(q)(6)(v)(E)** Know of the state emergency response plan and of the Federal Regional Response Team.
- **(q)(6)(v)(F)** Know and understand the importance of decontamination procedures.
- **(q)(7)**
- *Trainers.* Trainers who teach any of the above training subjects shall have satisfactorily completed a training course for teaching the subjects they are expected to teach, such as the courses offered by the U.S. National Fire Academy, or they shall have the training and/

or academic credentials and instructional experience necessary to demonstrate competent instructional skills and a good command of the subject matter of the courses they are to teach.

- **(q)(8)**
- Refresher training.
- **(q)(8)(i)**
- Those employees who are trained in accordance with paragraph (q)(6) of this section shall receive annual refresher training of sufficient content and duration to maintain their competencies, or shall demonstrate competency in those areas at least yearly.
- **(q)(8)(ii)**
- A statement shall be made of the training or competency, and if a statement of competency is made, the employer shall keep a record of the methodology used to demonstrate competency.
- **(q)(9)**
- Medical surveillance and consultation.
- **(q)(9)(i)**
- Members of an organized and designated HAZMAT team and hazardous materials specialist shall receive a baseline physical examination and be provided with medical surveillance as required in paragraph (f) of this section.
- **(q)(9)(ii)**
- Any emergency response employees who exhibit signs or symptoms which may have resulted from exposure to hazardous substances during the course of an emergency incident either immediately or subsequently, shall be provided with medical consultation as required in paragraph (f)(3)(ii) of this section.
- **(q)(10)**
- *Chemical protective clothing.* Chemical protective clothing and equipment to be used by organized and designated HAZMAT team members, or to be used by hazardous materials specialists, shall meet the requirements of paragraphs (g)(3) through (5) of this section (Figure 2.91).
- **(q)(11)**
- *Post-emergency response operations.* Upon completion of the emergency response, if it is determined that it is necessary to remove hazardous substances, health hazards, and materials contaminated with them (such as contaminated soil or other elements of the natural environment) from the site of the incident, the employer conducting the clean-up shall comply with one of the following:
- **(q)(11)(i)**
- Meet all the requirements of paragraphs (b) through (o) of this section; or
- **(q)(11)(ii)**

Figure 2.91 Chemical protective clothing and equipment to be used by organized and designated HAZMAT team members. (Courtesy: Edmond, OK, Fire Department.)

Where the clean-up is done on plant property using plant or workplace employees, such employees shall have completed the training requirements of the following: 29 CFR 1910.38(a); 1910.134; 1910.1200, and other appropriate safety and health training made necessary by the tasks that they are expected to be performed such as personal protective equipment and decontamination procedures. All equipment to be used in the performance of the clean-up work shall be in serviceable condition and shall have been inspected prior to use (OSHA).

Hazardous Materials Containers

Highway Transportation Vehicles

Transportation vehicles used to transport hazardous materials have number designations and carry certain types of hazardous materials on a regular basis (Figure 2.92).

MC/DOT 306/406 tankers are atmospheric pressure containers and have little or no pressure involved from the product. **They are recognized by their elliptical shape.** Typically, they carry fuels such as gasoline, diesel fuel, and fuel oil. There will usually be a -digit identification number in the center of placards or on an orange rectangle near the placard on tank vehicles. The primary hazard of materials carried in these tankers is flammability (Figure 2.93).

MC/DOT 307/407 tankers are low pressure containers and may be insulated or uninsulated. Insulated tanks have a **horseshoe shape** when

Figure 2.92 Highway transportation vehicles used to transport hazardous materials have number designations and carry certain types of hazardous materials on a regular basis.

Figure 2.93 MC/DOT 306/406 tankers are atmospheric pressure containers and have little or no pressure involved from the product.

viewed from the rear. They carry a variety of hazardous materials including poisons, flammables, light corrosives, and **elevated temperature** materials such as asphalt. There will usually be a -digit identification number in the center of placards or in an orange rectangle near the placard on these tank vehicles. There may also be a class 9 miscellaneous hazardous materials placard or a diamond with the word **HOT** or just the word HOT somewhere on the tank on each of the four sides (Figure 2.94).

Figure 2.94 MC/DOT 307/407 tankers are low-pressure and may be insulated or uninsulated. Insulated tanks have a horseshoe shape when viewed from the rear.

MC/DOT 312/412 tankers are low pressure containers and carry very heavy liquids so the tank diameter is small. There are heavy reinforcing rings around the outside of the tank. **These tankers are used almost exclusively for corrosive liquids**, which include acids and bases. The tanks may, however, be placarded oxidizer, flammable, or corrosive. The shape of the tank may be the only hint that there is a corrosive material involved (Figure 2.95).

MC 331 tankers are high-pressure tanks carrying liquefied compressed gases which have high liquid-to-gas expansion ratios. The tanks have rounded ends with no insulation and appear to be large "bullets on

Figure 2.95 MC/DOT 312/412 tankers are low-pressure and carry very heavy liquids so the tank diameter is small. These tankers are used almost exclusively for corrosive liquids

Figure 2.96 MC 331 tankers are high-pressure tanks carrying liquefied compressed gases which have high liquid-to-gas expansion ratios., south of the derailment. (Courtesy Miamisburg OH Fire Department).

wheels." Most are flammable, but nonflammable and poisonous materials may also be transported in these tanks. These containers are under high pressure and may rupture violently under flame impingement, or if the container is damaged in an accident (Figure 2.96).

MC 338 tankers carry cryogenic liquids; gases that have been liquefied for transportation. The tanks may have a box on the back where the transfer pump is located. All 338 tankers will have a silvery colored series of fins and tubes underneath the tank. This is the **heat exchanger** and sets the 338 apart from other types of tankers. The liquids have temperatures below –130°F, and some as low as –452°F. These liquids have great liquid to gas expansion ratios. The tanks are insulated and may or may not be under pressure. The materials inside may be inert, flammable, oxidizer, or poison (Figure 2.97).

Figure 2.97 MC 338 tankers carry cryogenic liquids (extremely cold); gases that have been liquefied for transportation.

Box trailers may carry any of the hazard classes of hazardous materials in almost any quantity. The placard requirements are based upon hazard and weight of the hazardous materials. Table 1 materials must be placarded regardless of quantity. Table 2 materials are placarded if there is 1,001 lb aggregate weight of two or more of the materials in the shipment. So if you have less than 1,001 lb of Table 2 materials, placards will not be required. For this reason, it is important to get an accurate count of unplacarded box trailers so that planners can estimate the number that may be hauling hazardous materials by using the national average percentages. It is estimated by the DOT that 50% of all highway vehicles are transporting hazardous materials. It is believed that only half of those are required to be placarded because of weight requirements.

The **Dangerous Placard** is used to identify mixed loads of hazardous materials and appear exclusively on highway box trailers (Figure 2.98). It may be used for mixed loads in which two or more Table 2 hazardous materials are in the shipment. When 5,000 lb or more of a Table 2 material are loaded in the shipment from one facility, there must be a placard posted for that shipment. The dangerous placard may still be used on the box trailer for two or more Table 2 materials that may be on the truck in addition to the 5,000 lb loaded at the one facility. A box trailer may have a number of different placarding configurations. The trailer may have a dangerous placard with other placards, it may have multiple placards from various hazard classes of Table 1 and Table 2 materials, or it may just have a single placard. The dangerous placard is not used in rail transportation.

Dry Bulk containers are used to transport dry materials such as flour and cement. Because the materials inside are very light, the container will

Figure 2.98 Dangerous placard is used to identify mixed loads of hazardous materials and appear exclusively on highway box trailers.

Figure 2.99 Dry bulk containers are used to transport dry materials such as flour and cement.

be very large compared to other highway tank vehicles. Very rarely will these containers be placarded (Figure 2.99).

Tube trailers are high-pressure tanks up to 3,000 psi. These tanks carry compressed gases. There are no liquids in the tanks. These gases may be flammable, nonflammable, oxidizers or poisons. The tanks can be recognized by the number of small individual tanks that appear to be stacked on a flatbed trailer and banded together (Figure 2.100).

FlatBed Trucks carry a wide variety of hazardous materials in packages and bulk containers. They may carry pallets of bags of oxidizers; 55 gallon drums of corrosives, flammable liquids, and poisons; and bulk containers of corrosives, flammable liquids, poisons, explosives, and flammable solids (Figure 2.101).

Figure 2.100 Tube trailers are high-pressure tanks up to 3,000 psi. These tanks carry compressed gases. There are no liquids in the tanks.

Figure 2.101 Flat bed trucks carry a wide variety of hazardous materials in packages and bulk containers.

Blasting agent vehicles are unique vehicles that carry both raw product hazardous materials and finished product blasting agent. The vehicles will have three placards: oxidizer, flammable liquid, and explosive 1.5, blasting agent. Fuel oil, flammable, ammonium nitrate, oxidizer, and a mixture of both can be found in three different compartments on the vehicle (Figure 2.102).

UPS/FED Express vehicles. Make note of all types of delivery vehicles such as Federal Express and United Parcel Service. These

Figure 2.102 Blasting agent vehicles are unique vehicles that carry both raw product hazardous materials and finished product blasting agent.

Figure 2.103 UPS/FedEx Vehicles. Make note of all types of delivery vehicles such as Federal Express and United Parcel Service.

vehicles may or may not have any placards on them, depending on the quantities carried. Those materials from Chart 16 that must always be placarded will require placards on these vehicles as well as others (Figure 2.103).

Rail Road Transportation

Railroad cars that are used to transport hazardous materials are, in many cases, similar to their highway counterparts, but they have much larger capacities. The primary types of railcars are box, hopper, bulk, tank, flat and tube. Types of hazardous materials carried by rail are much like those transported on highways, but again a primary difference is the increased amount of product being transported in individual containers. Additionally, there may be more than one tank car of the same product, or multiple products. Highway emergencies usually involve single amounts or small amounts of several chemicals, but rail incidents can involve from 1 to 100 or more tanks of the same or varying types of hazardous materials. Increased amounts of hazardous materials only compound the situation, making the work of response personnel difficult (Figure 2.104).

Railroad tank cars contain bulk quantities of hazardous materials and are the primary concern of emergency responders in derailments or other types of rail accidents. In fact, over 80% of all hazardous materials are transported by rail and tank cars account for 70% of the total. Most serious railroad hazmat incidents involve tank cars; therefore, they will be the focus of most of this section.

Figure 2.104 Railroad cars that are used to transport hazardous materials are, in many cases, similar to their highway counterparts, but they have much larger capacities.

Box Cars

Box cars are also used to transport hazardous materials. Individual container sizes inside box cars are 119 gallons or less, and in many cases 55 gallons or less. Potentially any class of hazardous materials may be included in box car shipments. Flat cars are used to ship pallets of hazardous materials including small containers. Flat cars are also used to carry intermodal containers. These can be box containers or any type of bulk tank containers. Intermodal containers get their name from the fact that they are shipped in highway, rail and water transportation. They can carry any class of hazardous material and the quantities will be smaller than ordinary highway or rail containers. There can, however, be multiple intermodal containers, which can present a quantity problem during an accident. Hopper cars do not always contain hazardous materials, but the physical state of the materials in the container may present a hazard. Fine powders and dusts when suspended in air can become a dust explosion hazard. An accident could cause the materials to be airborne, and if an ignition source is present, a dust explosion can occur.

Tank Cars

Cargoes Vary Greatly   More than 200,000 tank cars are used in rail transportation in the U.S. Most tank cars are not owned by railroads, but by private companies. Over 1,200 are owned by the U.S. government and supervised by the military. Government cars are used to transport jet fuels, nitrogen tetroxide (a rocket fuel), and materials used in weapons production. The amount of a hazardous material in a tank car can range from a few hundred gallons to as much as 34,500 gallons. (No tank car is considered empty unless it has been purged of the product and cleaned.)

Railroad tank cars are assigned identification numbers that identify the type of tank—pressure, nonpressure, cryogenic or miscellaneous.

Tank cars transporting hazardous materials can be identified through 100-series numbers regulated by the U.S. Department of Transportation (DOT). Tank cars that carry nonhazardous commodities such as corn syrup and cooking oil bear 200-series numbers and are regulated by the Association of American Railroads (AAR) are not considered hazardous materials containers.

Numerical designations for tank car designs are as follows:

- Pressure Tank Car Designations—DOT 105, 109, 112, 114 and 120
- Nonpressure Tank Car Designations—DOT 103, 104, 111 and 115; AAR 201, 203, 206 and 211
- Cryogenic Tank Car Designations—DOT 113; AAR 204 and 204XT
- Miscellaneous Tank Car Designations—DOT 106, 107 and 110; AAR 207 and 208

While the numbering system for railroad cars identifies particular types of tanks, the numbers may not be readily visible to emergency responders from a safe distance.

The most important concern for emergency responders about a tank car is whether it is a pressure or a nonpressure tank. Pressure tank cars present a much more difficult and dangerous situation for response personnel than nonpressure cars.

Pressure and nonpressure tank cars can be differentiated by looking at the locations of valves and other piping on the tops of the cars. On a pressure car, all valves and piping are enclosed within a dome at the top center of the tank car.

> **Hazmatology Point:** *An exception to this are cars carrying ethanol and crude oil. Incidents involving these materials have resulted in modifications to these tank cars to enclose the valves and other fixtures that might be broken off in a derailment within a dome cover. However, ethanol and crude oil are usually present in unit trains with large numbers of these liquid cars.*

The dome is designed to prevent damage during a derailment. Nonpressure tank cars can have unprotected valves and piping on top of the tanks or, in some cases, on top of the domes. Nonpressure tank cars may also have bottom fittings and washouts.

Figure 2.105 Pressure tanks are top loaded through the dome assembly and are generally used to transport flammable and nonflammable gases and poison gases.

Pressure Tanks   Pressure tanks are top loaded through the dome assembly and are generally used to transport flammable and nonflammable gases and poison gases. The tanks may be insulated or non-insulated and are hydrostatically tested for pressures of 100–600 psi. Pressure tank cars present a much more significant hazard to response personnel than do nonpressure tank cars under heat or flame impingement conditions (Figure 2.105).

Nonpressure Tanks   Nonpressure tank cars are hydrostatically tested from 35 to 100 psi. They are commonly used to transport flammable and combustible liquids, flammable solids, oxidizers, organic peroxides, poison liquids and corrosives. There are exceptions: for example, DOT 111 tank cars may carry specific flammable and nonflammable gases (Figure 2.106).

Tank cars are constructed from a variety of materials, including carbon steel, aluminum, stainless steel, nickel, chromium and iron. Single thicknesses of tank materials range from one-eighth to three-quarters of an inch. Tank car design standards are found in 49 CFR, Part 179 of the DOT Hazardous Materials Regulations. Modifications may occur to tanks that accommodate products transported because of product temperature, flammability or chemical reactivity. Tanks may be insulated externally to protect against the effects of ambient temperatures. Insulation materials can include fiberglass, polyurethane or pearlite; cork is used in some older tank cars.

Flammable pressure containers for transporting liquefied petroleum gases (LPGs) are double-layer tanks with an inside and outside tank (Figure 2.107). Tanks are then provided with thermal insulation between the tank layers designed to keep tank temperatures below 800°F during a 100-minute pool-fire or a 30-minute torch-impingement test (Figure 2.108).

Figure 2.106 Nonpressure tank cars are hydrostatically tested from 35 to 100 psi. They are commonly used to transport flammable and combustible liquids, flammable solids, oxidizers, organic peroxides, poison liquids and corrosives.

Figure 2.107 Flammable pressure containers for transporting Liquefied Petroleum Gases LPGs are double-layer tanks with an inside and outside tank.

This protection is provided by a layer of wool or ceramic fiber covered by a one-eighth-inch steel jacket. Thermal protection can also be provided through a textured coating sprayed onto a tank's outer surface. Heat from a flame exposure is absorbed by the coating material and is not transferred to the tank metal. Metal does not hold up well to direct flame contact and will fail in between 8–30 minutes, of the start of flame impingement on the vapor space of a tank. The average time of failure in 58% of the incidents is 15 minutes or less. according to information from the National Fire Protection Association (NFPA). Tank cars are protected from physical damage during an accident by shelf couplers and head shields. Shelf couplers are designed to stay together so they do not puncture another car

Figure 2.108 Tanks are then provided with thermal insulation between the tank layers designed to keep tank temperatures below 800°F during a 100-min poolfire or 30-minute torch-impingement test.

during a derailment or other accident (Figure 2.109). Head shields provide an extra layer of metal to help prevent dents and punctures when a pressure tank is hit by another car or object (Figure 2.110).

Cryogenic Tanks Cryogenic railroad tank cars are usually constructed of nickel or stainless steel as a tank within a tank. Cryogenic tanks may also

Figure 2.109 Tank cars are protected from physical damage during an accident by shelf couplers and head shields. Shelf couplers are designed to stay together so they do not puncture another car during a derailment or other accident.

Figure 2.110 Head shields provide an extra layer of metal to help prevent dents and punctures when a pressure tank is hit by another car or object.

be found within boxcars. Because cryogenic liquids are very cold, insulation is placed between the two tanks, and a vacuum is pulled on the space to maintain the temperature (Figure 2.111). This process will allow the tank car a 30-day holding time. Cryogenic cars transport various gases, including flammable hydrogen, liquid oxygen and poisons. Some cryogenic gases, such as nitrogen and argon, are considered inert. Temperatures of these liquefied gases can range from the warmest, carbon dioxide at –130°F, to the coldest, helium at –452°F. Thermal hazards of these materials are significant. In addition to thermal hazards, cryogenic liquids have a large liquid-to-vapor expansion ratio. A small leak from a valve or container can create a large vapor cloud. Some ratios are as great as 900:1, meaning 1 gallon of cryogenic liquid can produce over 900 gallons of gas.

Multiunit railcars are used to transport individual tanks of gases in uninsulated "ton containers." They are removed from the car for use, refilled and re-transported. Products carried include chlorine, phosgene,

Figure 2.111 Because cryogenic liquids are very cold, insulation is placed between the two tanks, and a vacuum pulled on the space to maintain the temperature.

anhydrous ammonia and refrigerants. "Ton containers" have a 180- to 320-gallon water capacity and are pressure tested from 500 to 1,000 psi. They may be transported by rail or truck and can be found on special flat cars, boxcars or gondola cars, as well as on "trailer on flat car" (TOFC) or "container on flat car" (COFC) units.

High-pressure tank cars (similar to highway tube trailers) are approximately 40 ft long and contain a series of 25–30 steel cylinders or individual tanks that are tested to 4,000 psi. High-pressure tanks are not insulated and are equipped with pressure relief valves, and are usually used to transport helium or hydrogen.

Another type of pressure tank car is only pressurized during unloading. This car is a pneumatically unloaded, covered hopper car. Pressure is applied during the unloading process, and the tank is tested to between 20 and 80 psi. This type of tank is used for dry caustic soda (*Firehouse Magazine*).

Warnings Posted   Railroad tank cars, as previously mentioned, are built to DOT specifications for hazardous materials transportation. Certain markings are required to be stenciled on each tank car as a part of the specification requirements. Of particular interest to emergency responders is the requirement that names of certain commodities be stenciled on both sides of the tank car in 4-in.-high letters. Some 50 materials require name stenciling; these include anhydrous ammonia, chlorine and liquefied petroleum gas. Additionally, if a material is an "Inhalation Hazard," that too must be stenciled on the container above the commodity name.

A tank containing hazardous materials will bear a DOT specification including DOT followed by the tank car type, such as 111 or DOT-111. Next to the tank car type will be a letter designating the type of protection the pressure car has for accidents or flame exposure:

- A—the tank has top and bottom shelf couplers.
- S—the tank has A plus head puncture resistance.
- J—the tank has A and S and jacket thermal protection.
- T—the tank has A, S, J and spray-on thermal protection.

If the tank is a pressure container, following the letters will be a number indicating the tank test pressure. Following the test pressure will be letters designating the type of material in which the tank is constructed: Al designates aluminum, N nickel and C, D and E for stainless steel. Nonpressure tanks will not have the test pressure information.

Unless they are carrying certain poison gases, such as hydrocyanic acid, pressure tanks have relief valves designed to relieve excess pressure caused by increases in ambient temperature. These relief valves are not designed to relieve the pressure created from radiant heat from a fire or other source or direct flame impingement (DOT).

Many accidents have occurred over the years from boiling liquid expanding vapor explosions (BLEVEs) when pressure containers have had excess pressure buildup. These accidents have resulted in many fire-fighter and civilian deaths and injuries when pressure relief valves could not keep up with internal pressure buildup and the tanks failed.

NFPA 472, 1997 Edition lists the competencies that a hazardous materials technician with a tank car specialty should have.

Water Transportation

Shipping containers may be placarded, and the placards could be visible if the containers are on the deck of the ship or barge. Barges do not have placards, but do have a red pennant or flag that indicates that the barge is carrying hazardous materials (Figure 2.112). The pennant does not identify the hazard class or type of hazardous material. Make note of liquid and pressure barges. Hazardous materials can be found on navigable rivers, ports and the open seas. Many times though the containers are below deck and not visible. It is possible to determine if barges are carrying liquids, solids or gases. The best bet for obtaining information about water shipments is to visit the port facility and identify the types of containers

Figure 2.112 Barges do not have placards, but do have a red pennant or flag that indicates that the barge is carrying hazardous materials.

Figure 2.113 The best bet for obtaining information about water shipments is to visit the port facility and identify the types of containers and placards shown on those containers sitting on the dock.

and placards shown on those containers sitting on the dock (Figure 2.113). Ocean-going vessels carrying hazardous materials are either container ships or tankers (Figure 2.114). You may also be able to obtain information from the shipping company.

Intermodal Containers

The use of intermodal containers is referred to by the DOT as "intermodalism," which means "the use of more than one form of transportation." Intermodal containers resemble many of the other bulk transportation containers found in highway, rail and water modes of transportation, but are smaller in capacity (Figure 2.115). They are designed to be transported in rail, water and highway modes without the need to off-load the products; that is, the container is transferred from one mode of transportation to another rather than the product being transferred from one container to another. When aboard a ship, intermodal containers may be found above or below the deck. Once a container ship reaches port, the intermodal containers are unloaded by giant cranes. The containers may also be shipped on inland waterways aboard barges. Containers are moved around on land by specially designed forklift-type vehicles that are much larger than a typical forklift. Containers may be found stacked on top of each other on

Figure 2.114 Ocean-going vessels carrying hazardous materials are either container ships or tankers.

Figure 2.115 Intermodal containers resemble many of the other bulk transportation containers found in highway, rail and water modes of transportation, but are smaller in capacity.

the dock awaiting further shipment on land. From the dock, the containers are moved inland by rail or truck.

Intermodal containers are not usually transferred directly from ship to rail or truck. They are unloaded onto the dock and then transferred to rail or truck by specially designed cranes. An added advantage to intermodal containers is that they are portable and can be taken to the end-use site and off-loaded as a stationary storage tank until the product is used up and then returned for refilling. Intermodal containers are regulated by the DOT, and containers are made to specifications prepared by the International Maritime Organization (IMO), an agency of the United Nations that deals with treaties for maritime safety among other matters. Primarily IMO Specification 1, 2, 5 and 7 tank containers are used to transport hazardous materials.

Intermodal containers come in many sizes and shapes. Box containers may be found in varying lengths from 10 to 48 ft. Typical heights are 8 ft, 6 in., but some styles may vary from 8 to 9 ft, 6 in. The standard width of intermodal containers is 8 ft. Containers may have doors at the rear with some also having side doors.

Intermodal containers may be insulated or uninsulated, and may have environmental temperature controls. Bulk tanks are designed with a steel structure around the outside to facilitate stacking and movement. They are usually 20–28 ft in length and are used for liquids and bulk materials. Tank containers have a capacity that ranges from 4,000 to 6,000 gallons.

Tank containers are used for transport of liquids or gases, including flammables, toxic and corrosive chemicals, cryogenic liquids, and others. Tanks are constructed of metal with two basic components: the tank itself and an outer framework. Usually, containers are constructed of stainless steel, but also aluminum, mild steel or magnesium alloy. Containers may be lined, refrigerated, heated with electricity or steam and insulated with metal or plastic jackets. Weight, volume and construction details of a tank container vary considerably due to the properties of the transported substance.

The IMO defines five different types of tank containers where the following two types are significant for chemicals. Within the United States, IM 101 and IM 102 portable tanks are the most commonly used containers for both hazardous and nonhazardous materials. They are equivalent to IMO 1 and 2 tanks. Specification 51 tanks in the United States are equivalent to IMO 5 tanks and are pressurized between 100 and 500 psi.

IMO 1 tanks for the highly flammable, toxic and corrosive liquids:

- Shell constructed of 316 stainless steel
- Capacity of 3,158–6,974 gallons
- Maximum Allowable Working Pressure (MAWP) of 58 psi (25.4–100 psi U.S.)
- For transport of hazardous liquids
- Available in 7.65-, 10-, 20- or 30-ft sizes
- Can be heated by steam or electric (glycol cooling optional)
- Flashpoints below 32°F

IMO 2 tanks for medium-hazard products such as flammable liquids, herbicides, resins and insecticides:

- Shell constructed of 316 stainless steel
- Capacity 3,158–6,974 gallons
- MAWP of approximately 25 psi (14.5–25.4 psi U.S.)
- For transport of low-hazard liquids
- Available in 7.65-, 10-, 20- and 30-ft sizes
- Heating by steam or electric and cooling by glycol
- Flashpoints between 32°F and 142°F

IMO 5 tanks for high-hazard products such as flammable gases:

- Shell constructed from 316 stainless steel
- Capacity of 3,846–6,710 gallons
- For transport of gases, including propane, butane, anhydrous ammonia and refrigerants
- Available in 20-, 30- or 40-ft sizes
- Cooling by sun-shield

IMO 7 tanks for cryogenic/refrigerated liquids:

- Capacity of 2,105–5,263 gallons
- Bulk liquids—oxygen, nitrogen, carbon dioxide, argon, ethylene

Some intermodal containers are specially designed to carry specific hazardous materials. These include the Type 7 for cryogenic liquids, sometimes referred to as refrigerated liquids, and tube modules used for compressed gasses. Tube modules are used for helium, nitrogen, oxygen and others. Pressures range from 3,000 to 5,000 psi.

Intermodal containers are marked with placards in the same manner as their full-size highway and railroad counterparts. The only exception is that they are placarded according to highway placarding regulations rather than railroad. On the railroad, all rail cars are required to be placarded regardless of the amount of product, so even if they are being shipped by rail, intermodal containers will be placarded according to the DOT's Tables 1 and Table 2 placarding requirements (IMO).

In addition, since these are bulk containers, a United Nations four-digit identification number will be assigned to the product and displayed in the center of any placards present or in an orange rectangle on the container. Additionally, international shipments have markings associated with the Agreement concerning the International Carriage of Dangerous Goods by Road (ADR). An additional orange placard will be on the container above the UN four-digit number with the ADR Hazard Identification Number (HIN). This is also known as the "Kemler Code."

Shown on the right is the combination of the UN four-digit number and the HIN for acetyl chloride. The number 1717 is the UN identification number and the X338 is the HIN. The first figure in the Kemler Code indicates the primary hazard (see chart above). In this case, the number is 3, which indicates a flammable liquid as being the primary hazard. The second two numbers are the secondary hazards also taken from the chart for number combinations (at right).

When a number is doubled, it indicates that the hazard is intensified for that particular hazard. Because there are two 3s, it is an indication that there is an extra fire hazard, which in this case is a self-heating material. The number 38 indicates that this material is a "self-heating liquid, corrosive." This particular chemical also has an X in front of the numbers, which indicates that this material is water reactive. Therefore, the X338 above the UN number tells you that acetyl chloride is a flammable liquid that is self-heating, and it is also corrosive and water reactive. Information concerning the ADR Hazard Identification Number can be found in the *Emergency Response Guidebook*.

Box-type intermodal containers are also a concern to national security as they may be used by terrorists to smuggle personnel and or weapons of mass destruction into the country. Security at ports is considered to be one of our vulnerable points. Congress is working with the Department of Homeland Security and the Coast Guard to beef up security at the nation's ports. People who work at port facilities and those who operate pleasure boats in the vicinity should be alert to unusual activity, which may be associated with potential terrorist actions. The Coast Guard should be notified if anything unusual is observed (DOT).

Fixed Facilities

The U.S. Department of Transportation DOT regulates the design and construction of shipping containers to help assure the safe transportation of hazardous materials. Specifications for fixed-facility containers, however, are quite different. There is no mandatory regulation of fixed-facility containers. Recommendations for fixed tank specifications are issued by codes and standards organizations. I have witnessed a railcar being used as an underground storage tank for a gasoline service station and highway transportation containers being used for fixed storage after they were no longer certified for highway use by the DOT. Fixed-facility containers can be of almost any size and shape. Some of them have names based on their designs and functions, such as the open floating roof tank. Others, however, do not have any specific name or designation, unlike their highway counterparts.

Several National Fire Protection (NFPA) standards address the storage of hazardous materials in fixed containers. These include NFPA 30 Flammable and Combustible Liquids Code and NFPA 58 Storage and Handling of Liquefied Petroleum Gases. Both of these codes address the design and safety requirements for specific fixed tanks such as propane and flammable liquid storage. The American Petroleum Institute, the Institute of Petroleum, the American National Standards Institute and others also publish standards—but like the NPPA codes, these are "consensus standards" and do not become law or mandatory until a jurisdiction adopts them. Some jurisdictions have passed local codes and ordinances that regulate certain hazardous materials in fixed containers, but these vary from location to location and do not have the consistence the DOT regulations provide for transportation of hazardous materials in containers.

Bulk Petroleum Storage

The first tanks to be discussed are those used for bulk petroleum storage. These include tanks with cone roofs, floating roofs, open floating roofs and retrofitted floating roofs. Many of these tanks are associated with tank storage facilities or "farms" where there are multiple tanks at one facility. Tank farms are often connected with pipelines as well as highway, rail and waterway transportation. Fixed storage tanks have pressures that range from atmospheric (0–5 psi) and low pressure (5–100 psi) to high pressure (100–3,000 psi) and ultra-high pressure (above 3,000 psi). Bulk petroleum tanks are generally considered atmospheric pressure tanks.

Cone-Roof Tanks

Cone-roof tanks get their name from the inverted cone-shaped construction of their roofs (Figure 2.116). They are atmospheric and low-pressure tanks with cylindrical outer walls supporting the cone roof. American Petroleum Institute Specification 650 calls for the roof-to-shell seam for this tank to be designed to fail as the result of a fire or explosion, reducing the possibility of a pressure buildup. Cone-roof tanks may be used to store gasoline, diesel fuel and corrosive liquids. Some of these chemicals, such as gasoline, are very volatile and easily produce vapor at normal atmospheric temperatures. Because a cone-roof tank is open inside, the surface of the liquid is exposed to air and product will be lost to vaporization. Therefore, a cone-roof tank is primarily used for nonvolatile materials. Contents of cone-roof tanks change frequently: daily, even hourly under some circumstances. Cone-roof tanks are recognized by the cone-shaped roof and the lack of wind girders or external vents around the top of the sidewall.

Figure 2.116 Cone-roof tanks get their name from the inverted cone-shaped construction of their roofs.

Open Floating Roof Tanks

An open floating roof tank has a roof that literally floats on the liquid product in the tank. The outer walls of these tanks are vertical and cylindrical, with the floating roofs eliminating the vapor space in the tank (Figure 2.117). Drains in place are designed to remove accumulations of moisture on the surface of the tank. An articulating arm-type device on top of the tank functions as a ladder for examination of the interior. As the product level and floating roof go down, so does the end of the arm, which rests on top of the roof. The location of this arm, when viewed from the outside, can be an indicator of the approximate amount of product in the tank. If the arm is completely visible, the tank is near full. If the arm is completely inside the tank, it is near empty. A wind girder is located near the top and on the outside of the tank. Primarily a retaining band, the wind girder provides necessary rigidity to the container wall when liquid levels are lowered. An external staircase is also located on the tank wall to provide access to the articulating arm (walkway) on top of the floating roof.

Closed Floating Roof Tanks

Another version of the floating roof tank is the covered or internal floating roof. These tanks may resemble cone-roof tanks but are distinguishable by the vents around the tank near the roof to the sidewall seam (Figure 2.118). The cone roof provides protection from rain and snow, while the internal floating roof eliminates the vapor space above the liquid and helps prevent

Figure 2.117 An open floating roof tank has a roof that literally floats on the liquid product in the tank. The outer walls of these tanks are vertical and cylindrical, with the floating roofs eliminating the vapor space in the tank.

Figure 2.118 Another version of the floating roof tank is the covered or internal floating roof. These tanks may resemble cone-roof tanks but are distinguishable by the vents around the tank near the roof to sidewall seam.

the loss of vapor into the air. This style of tank is sometimes used for polar solvent materials, which are miscible with water. If water or snow entered the tank, it would dilute the product. Materials stored in the internal floating roof tank are generally flammable and combustible liquids because of the vapor loss protection of the floating roof on top of the liquid. A floating roof tank may also be covered by a geodesic dome that keeps outside weather from affecting the floating roof in the tank. This form of internal floating roof tank is also used for flammable and combustible liquids.

Horizontal Tanks

Horizontal tanks are another kind of tank used for chemical storage in many parts of the country, particularly in rural areas. These tanks range in size from hundreds of gallons to thousands of gallons with low or atmospheric pressure (Figure 2.119). The ends of the tanks are flat, which usually indicates a limited amount of pressure. Many horizontal tanks are supported on stands constructed of metal. Unprotected metal can be dangerous during fire conditions, as the metal will be stressed by heat and fail very quickly. In the 1950s, six Kansas City, KS, firefighters were killed when steel supports of a horizontal tank at a gasoline service station failed,

Figure 2.119 Horizontal tanks are another kind of tank used for chemical storage in many parts of the country, particularly in rural areas. These tanks range in size from hundreds of gallons to thousands of gallons with low or atmospheric pressure.

Figure 2.120 In the 1950s, six Kansas City, KS, firefighters were killed when a horizontal tank failed sending burning gasoline into their location on the street. (Courtesy of Kansas City, MO, Fire Department.)

sending burning gasoline into their location on the street (Figure 2.120). NFPA 30 Flammable and Combustible Liquids Code requires that the steel supports of horizontal tanks be protected from flame impingement by encasement in concrete or other nonflammable material. Horizontal tanks may be used to store flammable and combustible liquids, corrosive liquids and many other types of hazardous materials.

A direct result of the Kansas City fire was the requirement for gasoline tanks at service stations frequented by the public to have underground storage tanks. According to NFPA, there has never been a fire or explosion involving an underground fuel storage tank. During the 1970s and 1980s, it was discovered that underground storage tanks, which were usually constructed of steel, eventually corrode and leak fuel into the ground. Gasoline and other flammable fuels have a specific gravity less than water, so they float on the surface of water. Moisture that accumulates inside underground storage tanks settles to the bottom of the tank and causes the metal to corrode.

Because of the leakage, the U.S. Environmental Protection Agency (EPA) requires that all underground tanks be replaced and safeguards put in place to prevent and detect leaks. An alternative to the placement of tanks underground was allowed in the form of an aboveground vault. The vault is constructed of concrete, and the tank(s) are placed within the vault, partially underground with the vault top above the ground. There is access to the vault from aboveground. Vaults for flammable and combustible liquids must have safety precautions such as monitors installed and must be able to hold the contents of the largest tank if it should leak.

Figure 2.121 Spherical tanks are often found in or near refineries and hold hydrogen gas. These tanks are not insulated, and flame impingement on the container surface can cause a violent rupture.

High-Pressure Tanks

Spherical tanks are often found in or near refineries and hold hydrogen gas. These tanks are not insulated, and flame impingement on the container surface can cause a violent rupture (Figure 2.121).

Propane and anhydrous ammonia are stored in high-pressure tanks that contain many safety features that are built-in through codes and standards. Industry associations are also heavily involved in safe storage and handling requirements for their members and emergency responders. Propane tanks range in size from 5-gallon containers used with barbecue grills and 250-gallon tanks used for home heating to bulk storage tanks containing thousands of gallons of product at propane facilities.

Propane tanks generally have rounded ends, which is a primary indictor of a pressure vessel. Pressure tanks are equipped with relief valves to vent excess pressure caused by increases in ambient temperature (Figure 2.122). Propane tanks have extensions on the relief valves to extend vapors well above the tank in the event the vapors catch fire. The normal relief valve height is between 6 and 10 in. This height allows ignited vapors to lie on the vapor space of the tank, causing it to fail quickly.

Anhydrous ammonia tanks do not have the extension, so if you see a horizontal pressure tank with tall relief valves, it is likely to contain

Figure 2.122 Propane tanks generally have rounded ends, which is a primary indictor of a pressure vessel. These tanks are equipped with relief valves to vent excess pressure caused by increases in ambient temperature.

propane; if the relief valves are near the tank surface, it is likely to contain anhydrous ammonia (Figure 2.123). Propane, anhydrous ammonia and other petroleum gases are liquefied under pressure before being placed in these tanks. The liquid level in the tanks is generally 80% of the tanks' capacity when full to allow for a vapor space. Pressure containers can be dangerous under fire conditions. Relief valves are designed only to relieve normal pressure increases, not those caused by flame impingement from a fire or radiant heat sources. Flame impingement on the vapor space of the tank can cause the metal to fail, and a boiling liquid expanding vapor explosion BLEVE (Figure 2.124) can occur within 8 to 30 minutes of the start of the flame impingement. On average, 58% of all BLEVEs occur in less than 15 minutes. The liquid level of the tank will absorb head from flame impingement and not fail as long as the liquid is in the tank and the impingement is below the liquid level of the tank. The flame impingement on the liquid level causes the propane in the tank to boil faster and produce more vapors. If the vapor is produced faster than the relief valve can vent it, tank failure can also occur. Cylindrical tanks are another type of high-pressure tanks, which may be found at refineries or other locations. These tanks look like giant balls suspended in a steel support structure. Tanks of this type can be used for propane and other liquefied petroleum gases, natural gas and hydrogen.

Figure 2.123 Anhydrous ammonia tanks do not have an extension on the relief valve, so if you see a horizontal pressure tank with tall relief valves, it is likely to contain propane; if the relief valves are near the tank surface, it is likely to contain anhydrous ammonia.

Figure 2.124 Flame impingement on the vapor space of the tank can cause the metal to fail and a BLEVE can occur within 8 to 30 minutes of the start of the flame impingement.

Tube Banks

Tube banks are a type of ultra-high pressure tank like there highway counterparts that are often found at compressed gas companies and distribution facilities (Figure 2.125). The pressures in this type of tank can be in excess of 3,000 psi. Tube banks are actually a series of individual tanks stacked together with valves and piping into a single outlet/inlet, similar to a fire department cascade system. Unlike the liquefied gases in high-pressure tanks, the materials in tube bands are gases. There is no liquid level or space to absorb heat. Under fire conditions, these tanks can fail very quickly.

Vertical Cryogenic Tanks

Vertical cryogenic storage tanks are often found next to manufacturing buildings, hospitals, bottled gas facilities and welding supply houses (Figure 2.126). A heat exchanger next to the tank or tanks is confirmation that these are indeed cryogenic tanks. Heat exchangers are a series of silver-colored tubes with fins next to the tanks. The heat exchanger allows for the cold liquids to be turned back into gases for use. Ambient air around the tubes and fins of the heat exchanger warms the cold cryogenic liquids into their gas state. Cryogenic tanks have narrow circumferences and are very tall. These are high-pressure tanks used to store cryogenic liquids, which are very cold. Cryogenics have boiling points of –130°C to –452°C. Some tanks, particularly those found at cryogenic production facilities, may each hold as much as 400,000 gallons. Types of

Figure 2.125 Tube banks are a type of ultra-high pressure tank like there highway counterparts that are often found at compressed-gas companies and distribution facilities.

Figure 2.126 Vertical cryogenic storage tanks are often found next to manufacturing buildings, hospitals, bottled gas facilities and welding supply houses.

materials found in cryogenic containers include natural gas, argon, nitrogen, chlorine and oxygen. They can be flammable, oxidizers or poisons.

Portable Containers

We will look at some of the different types of portable containers which may be found in transportation, storage and use. Portable containers are those used to hold hazardous materials in small quantities, which are easily moved from one location to another (Figure 2.127). Portable containers are constructed of many different materials including glass, aluminum, stainless steel, steel, plastic, wood, cardboard, lead and others. Portable containers are regulated by the DOT in transportation and governed by ASME, NFPA and OSHA standards in fixed storage and use.

However, portable containers can be made of any material that will hold the contents and may not be specification containers. In some cases, containers may be reused from some other purpose, which may create an unsafe storage condition. Portable containers include, but are not limited to the following types: wooden and fiberboard boxes, metal drums, fiberboard drums, plastic pails, glass carboys in protective containers, cylinders, ton cylinders, mailing tubes, special lead containers for radioactive materials, plastic and multiwall paper bags. Liquid quantities can range from a few ounces to several hundred gallons (Figure 2.127A).

Figure 2.127 Portable containers are those used to hold hazardous materials in small quantities, which are easily moved from one location to another.

Figure 2.127A Liquid quantities can range from a few ounces to several hundred gallons.

Dry materials may range from a few ounces to several hundred pounds. Gases generally do not have much weight but require a substantial container to hold the high pressures causing the tank to be very heavy.

Fiberboard drums (Figure 2.127B) and multiwall paper bags are used for dry materials such as ammonium nitrate fertilizer and calcium hypochlorite, both of which are oxidizers. If the containers become wet during transportation or storage, or during firefighting operations, the packaging material can become impregnated with the oxidizer once the moisture

Figure 2.127B Fiberboard drums.

evaporates. If the packaging materials are then exposed to heat or flame in a fire, they can burn vigorously because of the oxidizer imbedded in the container material.

Glass and plastic carboys as well as stainless steel kegs, 55-gallon drums, pint and gallon glass bottles, 5-gallon plastic pails and lined drums are used to transport and store many different types of acids and bases and a variety of other hazardous materials (Figure 2.128). Acids can react quickly with their container causing container failure and subsequent spill of material if incompatible construction materials are used. Not all acids and bases can be placed in the same types of containers. Damage can occur to a container if different acids or bases are mixed in a container. Plastic and glass bottles, 55-gallon drums and plastic and metal pails may also be used for flammable, poisonous or oxidizing liquids (Figure 2.129). Glass and plastic bottles placed inside lead containers are used for some radioactive isotopes (Figure 2.130). More substantial lead containers are used for high-level radioactive materials because of the shielding required to keep the radioactivity from leaving the container.

Portable container pressures range from atmospheric in the case of drums and bags to ultra-high pressures of 6,000 psi for cylinders. Nearly every type of hazardous material found in bulk quantities may also be found in small portable containers. Additionally, there are many hazardous materials that are not shipped or stored in bulk quantities, but are rather usually found in small containers, such as certain explosives. Ammonium nitrate (ANFO), dynamite, blasting caps, detonation cord, fireworks and other explosives are packaged and shipped in cardboard boxes (Figure 2.131).

Figure 2.128 Stainless steel kegs are typically used for corrosive materials such as nitric acid, which is also an oxidizer.

Figure 2.129 Plastic and glass bottles, 55-gallon drums and plastic and metal pails may also be used for flammable, poisonous or oxidizing liquids.

Quantities of hazardous materials in portable containers are sometimes so small that the DOT does not consider them a serious hazard and does not require placard or labeling. These materials are placed in a special class known at Other Regulated Materials (ORM-D). For example,

Figure 2.130 Glass and plastic bottles placed inside lead containers are used for some radioactive isotopes.

Figure 2.131 Ammonium nitrate (ANFO), dynamite, blasting caps, detonation cord, fireworks and other explosives are packaged and shipped in cardboard boxes.

charcoal lighter would be classified as a flammable liquid in bulk quantities; however, in quart cans for consumer use, it is an ORM-D. Oftentimes there may be large numbers of small containers in a shipment or in fixed storage, which collectively can create a serious hazard even though the individual container quantity is small.

Cryogenic hazardous materials, those with boiling points below –130°F, are shipped, stored and used in two types of portable containers: Dewar and cylinder (Figure 2.132). Dewar containers are nonpressurized vacuum-jacketed flasks with a 5–20 L capacity, very much like a thermos bottle. Cylinder containers are insulated with a separate vacuum jacket and a 100–200 L capacity. Cryogenic containers are usually found at gas bottling plants, college and university research facilities, private sector research facilities, and are replacing the oxygen gas cylinder in welding operations in many areas.

Figure 2.132 Cryogenic hazardous materials, those with boiling points below –130°F, are shipped, stored and used in Dewar containers.

Common cryogenic hazardous materials include oxygen, nitrogen, argon, hydrogen, natural gas, helium and others. Cryogenic liquids have very low boiling points and high liquid-to-vapor expansion ratios. Even a small Dewar or cylinder container could produce a significant vapor cloud during a release causing the displacement of oxygen in confined areas. While many of the cryogenic liquids are "inert" or nonhazardous, the vapor can still be an asphyxiation hazard when the oxygen in the air is displaced. In addition, the contents of cryogenic containers are extremely cold with temperatures between –130°F and –456°F below zero. Contact with these liquids could cause frostbite, solidification of body parts or both.

Cryogenic liquids and liquefied gases such as propane and butane do not have the same physical characteristics. While cryogenics are very cold, liquefied gases exist at whatever ambient temperature is surrounding the container. If the ambient temperature is 100°F, then the liquefied gas is also around 100°F. If the ambient temperature is –40°F, then the liquefied gas is also around that same temperature. Portable tanks of liquefied compressed gases can be found in many sizes and shapes (Figure 2.133). They can contain many types of gases, but commonly contain propane or other Liquefied Petroleum Gas (LPG) used for home barbecue grills,

Figure 2.133 Portable tanks of liquefied compressed gases can be found in many sizes and shapes.

heating living spaces (Figure 2.134), as a fuel for motorized equipment , soldering and others (Figure 2.135). These containers shouldn't be over-filled. There should be a vapor space when the container is filled, usu-ally around 20% of the volume of the container, to allow for expansion of gas within the container during increases in ambient temperature. While not as extensive as cryogenic liquids, liquefied petroleum gases also have

Figure 2.134 Containers can hold many types of gases, but commonly contain propane or other LPGs used for home heating living spaces.

Figure 2.135 LPG also used as a fuel for motorized equipment such as this forklift.

large liquid-to-gas expansion ratios in the area of 270:1, so even a small container can produce a large amount of flammable gas when released to the atmosphere.

Pressures in LPG containers can range from 15 to 230 psi depending on ambient temperature around the container. The higher the temperature outside the container, the higher the pressure inside the container will be. Container bursting pressures are generally about four times greater than the working pressures of LPG. Rapid heat buildup can still cause container rupture if the pressure cannot be relieved fast enough. LPGs are heavier than air and would be found in basements or other low-lying areas during a release. They have no natural odor and are colorless. An odorizer is added to allow leaks to be detected.

Compressed gas cylinders are also quite common and often used for oxygen, nitrogen, carbon dioxide, fire extinguishing agents, hydrogen and many others (Figure 2.136). These containers are constructed of heavy steel and have operating pressures of 3,000–6,000 psi. Just the pressure alone in the container can present a significant physical hazard if the valves are knocked off or the containers are exposed to high heat or direct flame contact. These containers can rocket a great distance and present an impact hazard to building occupants and response personnel.

Cylinders in storage should have a protective valve cap in place over the valve and be secured in place or to a cart. Exposed valves can be sheared off if a cylinder is knocked over. Cylinders in use should also be secured in place. Cylinders should never be moved or transported with the regulator in place; it should be replaced with the protective valve cap. Fire code inspectors should watch for the proper storage and use of high-pressure cylinders during inspections. While many compressed gas

Figure 2.136 Compressed gas cylinders are also quite common and often used for oxygen, nitrogen, carbon dioxide, fire extinguishing agents, hydrogen and many others.

Figure 2.137 Ton containers (2,000 lb capacity) are used most often for the storage of chlorine and sulfur dioxide gases.

cylinders are painted, there isn't any reliable color code system to identify the contents from the color of the container.

Ton containers (2,000 lb capacity) are used most often for the storage of chlorine and sulfur dioxide gases (Figure 2.137). These cylinders can be found in water treatment facilities, waste treatment plants and swimming pools. They are shipped on specially equipped rail and highway vehicles.

Figure 2.138 The acetylene tank is a small "chunky" tank specially constructed to contain this highly unstable gas.

Acetylene is a common flammable gas usually associated with welding operations in conjunction with gaseous or cryogenic oxygen. Acetylene is a highly flammable gas with a wide flammable range (2%–80%), which is quite unstable at elevated pressures above 15 psi. The acetylene tank is a small "chunky" tank specially constructed to contain this highly unstable gas (Figure 2.138). An inert material such as fuller's earth or lime silica are placed in the acetylene container to absorb acetone, a solvent used to dissolve the acetylene gas and maintain its stability. The acetone keeps the acetylene in suspension preventing accumulation of pockets of high-pressure gas, thus stabilizing the explosive tendencies of the gas. Acetylene tanks should never be stored or transported on their side as the acetylene gas can separate from the acetone producing a potentially explosive situation.

Because of their relatively small size, portable containers can be found in almost any type of transportation mode including private vehicles. They can also be found in most types of occupancies including residences and garages. The quantities of materials present don't create a high level of risk to the community. However, they certainly can be a hazard to response personnel, especially if we are not looking out for them. Knowing the design characteristics of hazardous materials, containers can assist response personnel in safely dealing with a hazardous materials incident involving a bulk container. They can more effectively identify the dangers of the hazardous material as well as provide effective damage assessment on the container (*Firehouse Magazine*).

Incident History Is a Part of the Standard of Care

Studying historical incidents can help responders learn what works and what has not worked in dealing with hazardous materials. There have been hundreds of incidents starting with the earliest I was able to find that occurred in Syracuse NY in 1841. During my research I documented over 800 pages of incidents, many more than I had room for in Hazmatology. Many more were discovered and read about, but did not document them. One of the keys to safe and effective and efficient hazmat response is looking at lessons learned from historical incidents and coupling that knowledge with what we know today. It is hoped that will keep us from repeating incidents in the past that killed responders and the public.

Historical Incidents 1800s

Baltimore, MD, November 30, 1846, Powder Mill Explosion

On Monday morning last, the city of Baltimore was startled by a sudden shock, which was followed in quick succession by two others, causing houses to tremble as by an earthquake. A cloud of smoke immediately arose in the north-west, and the alarm of fire was raised. The smoke indicated that some terrible catastrophe had occurred at the extensive Powder Mills at the junction of the Falls Turnpike and the Susquehanna Railroad, about seven miles from the city. All doubts were soon removed by the intelligence that one of the Mills had exploded, followed by a distressing loss of life, and great destruction to property.

The explosion took place in the graining mill, in which were five persons, all of whom were instantly killed. Their bodies, after removing the remains of the building, were found in quite a mutilated state. In short so general a destruction has seldom or never been visited from such a cause. The powder house is totally annihilated. The explosion is supposed to have been caused by the wind blowing sand among the powder, whilst being ground, by which friction was produced. The quantity of powder in the mill must have been great to have produced so tremendous an explosion, and to have done so much damage. Considering the extent of injury done to dwellings in the neighborhood, it is fortunate that no more lives were lost.

Republican Compiler, Gettysburg, Pennsylvania, **1846-11-30**

Wilmington, DE, August 3, 1855, Powder Mill Explosion

Three of Gareshe's powder mills near Wilmington, Del., exploded this morning, about 8 o'clock, with most disastrous consequences, six persons having been killed and five wounded. The drying house of Gareshe's powder works at Eden Park, which exploded this morning, contained about one and a half tons of powder, and the force of the explosion was tremendous. The house had been in operation for forty years without accident, and was considered the safest house in the works. All the men connected with its operations were killed at once, being blown to atoms and found at various distances from 50 to 200 yards from the scene of the disaster.

A 14-year-old boy who was riding past, was also among the killed.

The explosion of the drying house caused a mill, situated about 300 yards distant, also to explode. In this mill six men were dangerously hurt. One of them was thrown a distance of 40 feet, and it is feared will not recover, another who was running from the mill, had his skull fractured by a piece of lightning rod, and his recovery is also doubtful. The engineer says there were three separate explosions in the mill. To those at a distance it seemed like a single prolonged explosion.

There were about fourteen hundred pounds of powder in the mill, which was less than usual. Mr. Garfesche says the explosion must have been caused by one of the Frenchmen, who was addicted to smoking his pipe, although he had been forbidden to do so.

The New York Times, New York, **1855-08-04**

Washington, DC, June 18, 1864, Arsenal Explosion

The immediate vicinity of the accident, inside of the Arsenal grounds proper, the charred remains of those who had perished were laid upon the ground and covered over with canvass. Nine people died and seven were badly burned or otherwise injured and survived. The scenes while the fire was in progress was truly heart-rending. Those who could, jumped from the windows, and many of them fainted as soon as they alighted on the ground. By the heroism of some persons present, some of the girls who were enveloped in flames, were saved from a frightful death. One young lady ran out of the building with her dress all in flames, and was at once seized by a gentleman, who, in order to save her, plunged her into the river. He, however, burned his hands and arms badly in the effort. Three others, also in flames, started to run up the hill, and the upper part of their clothing was torn off by two gentlemen nearby, who thus probably saved the girls from a horrible death, but in the effort, they too were badly injured.

The Evening Star, Washington, DC, **1864-06-18**

Cincinnati, OH, July 5, 1864, Train Explosion

When approaching Seymour our train was signaled to stop, when news was brought to us that a powder magazine had exploded on the Aurora train, and that several men had been seriously hurt. The train from Aurora, when nearing the Seymour depot, was firing salutes from a cannon, which was attached to the last platform of the train. A small furnace, used for heating the rod with which the cannon was set off, lay close by the breach of the gun, and the wind, it is supposed, blew a spark of this fire into the chest where about a keg or two of powder had been emptied. An instantaneous explosion was the consequence, and six men who were on the car at the time were blown into the air and horribly mangled. We arrived on the ground some five or six minutes after the accident had occurred, and found the men in a state of terrible agony.

They had been picked up and conveyed to a house a short distance from the scene of the disaster, where they were speedily attended to and their wounds carefully dressed. We have seen may exhibitions of hospital practice, and many outside cases of mutilation and accidents "by flood and field," but we never in our lives saw anything to equal the appalling scene which was presented to us yesterday. The six victims of the unfortunate catastrophe were literally stripped naked, every particle of clothing having been torn from their bodies by the force of the explosion, their limbs were fractured, their bodies were mangled, the hair of their heads was burned off, and the groans and cries of the sufferers filled the house where they lay. Dr. Rogers, of Aurora, was the principal physician in attendance, and in justice to him we must say that he worked most untiringly to allay the sufferings of the wounded men.

New York Times, New York, NY, **5 Jul 1864**

Coralville, IA, July 1875, Paper Mill Explosion

"The Iowa City Disaster"

Terrific Explosion of Chemicals in a Paper Mill Six Men Killed--Wreck of an Important Manufacturing Concern. The explosion occurred in the tank room, tank No. 3, weighing 6,000 pounds, being lifted from its bottom and blown so high that it looked no larger than a flour barrel, and falling into the river. It was about 9 o'clock last evening in the dusk. The gang of hands that came on at noon were within three hours of the end of their stent. Frank Chiba, the fireman, whose duty it was to regulate the steam passing into the tanks, was at his post. Joseph Smally was in the stock room, behind tank No. 4. Tierney was at the straw cutter, and Gilmore and Sinton were in the room over the tanks and Herman Bechtel, an employee of the flouring mill nearby, was chatting with Chiba in the boiler room. Those nearer to the scene say that there were four distinct explosions, which was doubtless the case, as No.3 would naturally cause the explosion of 1, 2 and 4, following in rapid succession, yet at an interval noted by the ear. These six persons were instantly killed by the explosion.

The testimony of those who saw the explosion at a distance from a quarter to a half mile, is unanimous about a red glare being seen. Our opinion is that it was a chemical explosion, fortified by these premises: The straw is pulped by the use of chemicals, lime, and acids, muriatic and sulfuric. The immense force of the explosion could not have been gathered on a fifty-pound pressure. The Inquests were held on both of which a verdict was reached that the deceased persons came to their death by the explosion of bleaching tanks in the Coralville paper mill.

The Iowa City Press, IA, July 1875.

McCainsville, NJ, July 2, 1886, Atlantic Dynamite Co. Explosion

Ten men were launched into eternity this morning by the explosion of more than a ton of nitroglycerine on the outskirts of the little village of McCainsville, distance from this place about five miles. The men entered the mixing house of the Atlantic Dynamite Company's works at 7:20 o'clock. Ten minutes later there was a tremendous explosion. The shock was felt at a distance of 15 miles from the scene. The big mixing house was simply obliterated. The men who occupied it were rent limb from limb. Their bones were ground to powder. Infinitesimal and almost unrecognizable portions of their bodies were picked up a thousand yards away, and so thoroughly had the powerful explosive performed its work that after a day's search by hundreds of people the remains discovered filled but a small space in the single coffin that will be followed to the grave this morning. There were no wounded.

Five of the victims were married and five single. None of them had been in the employ of the Atlantic Dynamite Company less than five years, and all of them were considered industrious and careful men. They understood that they carried their lives in their hands, and knew by heart the rules governing the manufacture of the dangerous explosive which they daily handled in one shape or another. The company employs about 100 men. It owns about 400 acres of land in Roxbury Township, and upon it, placed at wide intervals, are about 60 buildings. Some of these are of brick, but most of them are frame. The nitroglycerine and mixing houses are a mile from any dwelling. The former is built in the midst of a wood, on a gentle slope. About 200 yards from it, at the foot of the slope, stood the mixing house. It was a strongly built frame building, 60 feet broad by 80 feet long. It was one story in height, and was painted a color not inappropriately known as hurricane blue. To the east, west, and south of the mixing house is a grove of young white birch, with here and there a willow. The ground is swampy. To the north is a large field, its green sod thickly studded with daisies and buttercups. In a southwesterly direction stands the acid house, and further west the soda house.

From the nitroglycerine to the mixing house ran a large wooden trench. It was rubber lined and covered with a wooden roof. Through this trench the nitro was carried to the mixing house, where it was poured into a leaden tank six feet in diameter and five feet deep. It was run from this tank to a tub in small quantities, and there measured before it was transferred to three huge saucers. Into these saucers the "dope," which consists of nitrate of soda, sawdust, wood pulp, and other ingredients, was shoveled. Then the dangerous mass was pulled from one side of the saucers to the other and stirred and kneaded with wooden hoes. All the tools used were of wood. The men who handled them wore ordinary working clothes and, so far as known, ordinary boots. They were cautioned against dropping any of their tools and against stamping their feet. In fact, they were instructed to handle everything in the building as if it were made of Venetian glass.

But two of the men employed in the mixing house this morning were mixers. The others were packers, but had been in the employ of the company so long that they had a general idea of every branch of the business. The company manufactures seven grades of giant powder, four grades of Judson powder, and two of Penniman powder. The capacity of the works is 500,000 pounds per month. There was an unusually large number of men employed in the mixing house this morning, as the company intended to close the works to-morrow and not to reopen them till Tuesday. There was enough material in the mixing house to make four tons of dynamite and near the entrance, on a little railroad, over which loaded cars were drawn to the storehouses by a mule, stood a car. It also contained a quantity of dynamite.

At 7 o'clock this morning, Alfred Lovell, the Superintendent of the works, reached the nitroglycerine house. The mixing house was then unoccupied. Twenty minutes later he saw the men enter it. King, the father of one of the men who was killed, also entered the building, but left it immediately, carrying in his hand a can which he had been ordered to fill with ammonia. The mixing house had disappeared.

When the nitroglycerine burst its bonds the atmospheric pressure on the acid house seemed to be from within. The windows were blown to atoms, and the walls bulged outward. The roof fell, but in falling the rafters formed an arch. From beneath this arch Epner crawled, while the air was full of acid. The latter scorched the ground as it fell, and its stifling fumes rose in clouds from the steaming earth while daylight lasted.

The force of the explosion was principally directed to the south and west, and in those directions it was checked by hilly ground. The woodwork of the mixing house was transformed to sawdust. Nothing remained of it to tell whether it had been made of wood, brick, or stone. One of the foundations, a boulder weighing 500 pounds, was buried a distance of 100 yards. Nothing remained of the car of dynamite.

Where the nitroglycerine tank had stood was an immense crater. Where the saucers had been were smaller craters. For a square of 100 yards, the ground was a succession of mounds and depressions, the latter resembling huge buffalo wallows. The cause of the explosion seemed to be at once divined by all who heard it and hundreds crowded to the spot. They noticed that the hot water pipes by which the building had been heated were untouched. A search for the remains was at once begun. Men, women, and children tramped through the woods and fields on their sorrowful quest. In an oak tree that grew 500 yards from the mixing house was found a portion of a trunk. A boy found a foot and piece of an ankle.

On the foot was a shoe that crumbled at a touch. Here and there, but always at a great distance from the scene of the explosion, was picked up a bit of charred humanity. The bits were generally so small that it was impossible to say with certainty of what portion of a body they had once belonged. The largest fragment discovered was the portion of a body they had once belonged. The largest fragment discovered was the portion of a trunk, to which a part of a thigh bone was attached. The remains were placed in a box which had once held dynamite cartridges. Identification was out of the question, and it was decided to enclose them to one coffin.

Undertaker James Jardine, of Succasunna, was instructed by the company to prepare them for burial. He obtained a rosewood casket, to which will be affixed a plate on which the names and ages of the ten men will be engraved. Superintendent Lovell and Chemist Penniman could not account for the accident. They were certain it was not due to carelessness, and hoped the word would not be used. The men were sober, careful, and industrious. They appreciated the dangerous nature of their work. It was simply one of those accidents which no amount of care could avert. They rather depreciated the force of the explosion, and seemed anxious to create the impression that it had been little felt in the other buildings of the company. Men were engaged this afternoon in replacing the glass which had been broken in the office. The latter is at least a half mile from the mixing house.

The New York Times, New York, NY, **3 Jul 1886**
Indianapolis Star, Indiana, **1911-06-27**

Rochester NY December 30, 1887, Naphtha Explosion

An accidental discharge of 14,000 gallons of naphtha in one of the main sewers in Rochester, N. Y., on Wednesday, produced the most sensational and uncommon disaster that city or any other ever experienced. Soon after three o'clock the heavy stone covering of a manhole of the Platt

street sewer was blown off by a terrific explosion, and the almost immediately another explosion occurred beneath the Clinton Flouring Mill on
Mill street. The upheavals were followed by sheets of flame that burst out
with great fury to a height of sixty feet. The Chuton mill took fire first,
and the flames spread quickly to the Washington mill and the Jefferson
mill. These three mills and contents were destroyed, involving a loss of
$200,000. The first explosion was followed quickly by others along Mill
and Platt streets, and at several points on West avenue making over forty
in all, and extending along four miles of sewers. In each case the man-
holes were all blown to pieces, and in many places flames shot up and
continued to burn fiercely for several minutes. The explosions were so
violent as to hurl rocks into buildings and against pedestrians, causing a
panic throughout the whole region traversed by the sewers.

Four persons were killed, three reported missing and twenty badly
injured. The cause of the peculiar disaster was an attempt to pump the
naphtha from the Vacuum Oil Company's tanks through a two-mile con-
duit to the Municipal Gas Company's works near the center of the city.
The naphtha became ignited and exploded.

The Cranbury Press, New Jersey, **1887-12-30**

Chicago IL, December 11, 1888, Oatmeal Mill Explosion

An Explosion, at first supposed to have been that of the boiler, shattered
the building at the corner of Halstead and Fulton street, occupied as an
oat-meal factory, at a few minutes before two o'clock this morning. The
shattered walls fell upon and crushed a saloon and the wagon factory
adjoining, and in a moment all three buildings were in flames. Four men
were known to have been in the oat-meal mill at the time of the explosion,
and all but one was instantly killed. The streets for a block in every direction were strewn with debris of every sort. There was not a whole pane
of glass in any building within a quarter of a mile of the mill. Lake Street
for a block east and west of Halsted street was sprinkled with the glass
of demolished windows. Here and there the panels of a door were blown
out. And this was one of the curious features of the accident. The glass
and woodwork were pushed out instead of smashed in, as one would naturally expect them to be.

What caused the explosion could not be ascertained. Several people
who claimed to have been familiar with the interior of the mill thought
it was a boiler explosion. Others were positive that it was an explosion of
oatmeal dust, and that no boiler could cause an explosion so terrific as to
cover the neighborhood several feet with debris. George Barber, the day
engineer, said that when he left the building at evening the engines and
boilers were in proper condition, and he did not think that the wreck was
caused by a boiler explosion. The losses will aggregate $150,000.

Eight horses were burned to death. By nine o'clock the fire department had succeeded in extinguishing the fire, and the work of excavating in the ruins was begun. Up to ten o'clock no bodies had been recovered. In prosecuting the excavation the boilers were found to be intact. This does away with the theory that the explosion was the result of the bursting boilers. About 10:30 o'clock the body of a man supposed to be that of Engineer Miller, was found in the north drive-way of the mill. It was covered with about three feet of debris. The lower part of the body was covered with bruises and burns and the face was so mangled that it could not be recognized. The unfortunate man had evidently been blown from the engine room clear to the drive-way.

The Rolla New Era, Missouri, **1888-12-22**

Louisville, KY, June 30, 1890, Standard Oil Refinery Explosion

Five acres of fire was the sight witnessed at the Standard Oil Refinery at Fifth and C Streets this morning. The immense structure was blazing at every point, and the heat was so intense that even 200 yards away persons were overcome. The following is a list of the casualties: It was first reported that seven had lost their lives and later that three were killed and two were injured.

The fire broke out at 8:45 o'clock and was in many respects a remarkable one. The refinery is on the east side of the Louisville and Nashville track, but the tanks are scattered along it on both sides. On last Saturday a tank of crude oil came in on a flatcar from Cleveland, and it was to be turned into the refinery vats. Some of the workmen thought the iron tank was too hot for such a thing to be done with safety. After consultation it was postponed in the hope that the weather to-day would be cooler. It did not prove to be however, and it became necessary to run the oil out of the car tank into another one in the yard. Workmen climbed on the car. They mounted the man head and were about to unscrew the cap, when they felt that there was a tremendous pressure from the inside. At first they decided not to open it, but finally they changed their minds and did so.

In an instant there was a dull puff and the vapors escaped, filling the air all around. The gas, as is known, is heavier than the air, and it sank to the ground, spreading out all over the locality, and moving with the wind. Almost in a twinkling it reached one of the sheds under which was a fire. There was a flash as the inflammable vapor ignited, and immediately after there was a tremendous explosion. The tank was blown to pieces, and the hundreds of gallons of burning oil were scattered all over the great works. A wall of fire 300 feet high and nearly 900 feet long moved with lightning rapidity to the buildings. In less time than it takes to relate it the canning house, filled with thousands of gallons of canned oil, the cooper shop, carpenter shop, pump and engine houses, the filling and lubricating houses, the storage houses, the paint and glue houses, and 900 feet of the platform were all burning furiously. The buildings and stock so far as burned will be a total loss. I should judge now, from present appearances, that the loss on stock and all would be between $30,000 and $40,000."

The New York Times, New York, NY, **1 Jul 1890**

Kings Station, OH, July 16, 1890, Railroad Car Explosion

A special from Cincinnati, O., says a terrible explosion occurred late yesterday afternoon at King's Station, on the Little Miami railroad, twenty-nine miles east of Cincinnati. Ten or more persons killed, and over thirty seriously injured. Two empty freight cars were being switched onto a side track where a car containing 500 kegs of gunpowder was standing. As the cars struck there was an explosion, and immediately afterward another car containing 800 kegs of gunpowder exploded. A brakeman in the employ of the Little Miami, was blown to atoms. Five other persons, supposed to be employees of the powder company, were killed.

The Kings Powder company and the Peters cartridge works have buildings on both sides of the river along the railroad. The explosion occurred on the south side and the destruction was enormous. A number of cottages occupied by workmen in the powder factory, and situated close to the track were shattered, and their inmates injured. Twelve or fifteen girls at work in the cartridge factory were crippled.

The railway station and freight house belonging to the Little Miami railroad, together with all the adjacent buildings, were set on fire and consumed. The track and ties of the railroad are fairly torn out of the ground and a great hole plowed in the earth. The Peters cartridge factory was burned to the ground. A relief train was dispatched from Cincinnati to the scene of the disaster and the work of rescue and relief was afforded as soon as possible. The work of searching for the missing and caring for the wounded is now progressing.

Newark Daily Advocate, Newark, OH, **16 Jul 1890**

Blandford, VA, April 7, 1894, Fireworks Factory Explosion

Fireworks blow up. Eleven men killed in an Explosion. Girls have a narrow escape. This afternoon about 3 o'clock, an alarm of fire was turned in and soon followed by a loud explosion, and in about fifteen minutes thereafter there was a second explosion. These explosions were distinctly heard for over a mile, and were caused by fire breaking out in the fireworks factory of C. N. Romaine & Bro., in Blandford. The fire originated in the building where powder for whistle bombs were made. The flames spread very rapidly, and from distant portions of the city great clouds of black smoke could be seen rising. The flames were quickly communicated to the other buildings, used for the manufacture of fireworks, and there were frequent small explosions. It was reported that there was a large quantity of powder stored somewhere exactly where no one appeared to know near these buildings, and this had the effect of keeping back a large proportion of the crowd from venturing too near the fire. On the opposite side of the street from the fireworks building, all of which were frame structures, was the trunk factory of Messrs. Romaine Bros., and close by was the large brick tobacco factory of Bland Bros. & Wright and the old whisky distillery, now unused.

All of these buildings, with the stock and machinery, were burned to the ground, as was also a large quantity of lumber. It was impossible to ascertain definitely what the loss by fire will be, but it is estimated that the total loss cannot be less than $75,000 or $100,000; partially covered by insurance. There were three explosions. The first was a small affair. As soon as it occurred, Messrs. Romaine, Bland and Tosh rushed into the drying room, and there the second and fatal explosion occurred, and they were killed. A number of girls employed in the fireworks factory escaped just before the second explosion.

Atlanta Constitution, Georgia, **1894-04-08**

Elliotsville, WV, November 30, 1889, Powder Explosion

Children annihilated. Four little girls blown to atoms while playing with powder in an abandoned mine. The mother driven insane when told of the terrible fate of her loved ones.

Four little girls, children of Hugh Dunn, a wealthy mine owner, found a keg of powder in an abandoned working Thursday morning. In some way they exploded it and were blown to atoms. Their mother lost her reason when told of the fate of her children.

Eau Claire Leader, Wisconsin, **1889-11-30**

Blue Island, IL, August 22, 1890, Standard Cartridge Factory Explosion

Six Men Instantly Killed and Thirteen Badly Injured Near Chicago. There was a terrific explosion in the Standard cartridge factory at Blue Island, twelve miles out of Chicago, Ill. The building was torn to pieces, and window glass was broken for a half mile around. The explosion occurred in the packing house. There was nobody at work in this portion of the factory, but the concussion destroyed the engineer's room, where several men were employed.

Six laborers were killed outright and thirteen others were injured, four of whom will die. The bodies of the dead were so badly mangled that identification was impossible. When the Rock Island passenger train rolled into the Lake Shore station that night it had a car full of men who were bleeding from wounds. They were taken to the county hospital in three ambulances. The fatally injured were sent to Chicago on the Rock Island train to be conveyed to the county hospital. One of the injured is reported to have died on the way to the hospital.

When help came to the unfortunates girls were found lying half naked in the mud and grass, and crying for someone to relieve their sufferings. Stripped of their clothing, they were blistered and blackened and swollen till their most intimate friends scarcely knew them. Of the twelve girls employed in the factory only three escaped without injury. The remaining nine were removed to a farm house half a mile away, where they were given medical attention, and afterwards supplied with clothing and removed to their homes. The ages of the girls ranged from 14 to 19 years.

A female worker was engaged in packing and labeling cartridges not far from the machine that exploded. A male worker who was one of the most terribly burned, was feeding blank shells into the tube of the machine that exploded. He was blown up in the air by the explosion and fell in the fire. There he lay, unable to extricate himself from the twisted rods, until the smoke had partially blown away and he was helped out by his brother. His face, arms and body were almost cooked in places, and every particle of clothing he had on save his shoes and stockings was burned off.

Another male worker whose injuries will probably cause his death, was working at a machine just next to the one that was blown up. Almost every portion of his body was burned, and he was gashed about the abdomen. His wounds were burned and blackened by the powder.

The proprietors of the factory say the disaster was caused by the gross carelessness of one of the men in opening a can of powder with a hammer and chisel.

Evening Gazette, Sterling Illinois, **1890-08-22**
The Cranbury Press, New Jersey, **1890-08-29**

Tarrytown, NY, May 22, 1891, Explosion and Train Wreck

A flat car loaded with dynamite and drawn by a construction engine was blown to atoms shortly before noon a few days ago, at a place about one and a quarter miles south of Tarrytown, N. Y., on the New York Central and Hudson River Railroad. A gang of workmen, chiefly Italians, were on the car, and of the number thirteen were killed, ten were injured and five, on the day after the tragedy, were missing. It was thought that the bodies of the latter were thrown into the river. Efforts were made to recover them. The bodies of several of the dead Italians have neither been claimed nor identified. Some of the injured of the same nativity have been recognized. Three of the wounded were taken to a local hospital at Tarrytown; the rest were brought to Bellevue Hospital in New York City, where they are being properly cared for.

Had the explosion occurred while a heavily loaded passenger train was passing the work train the loss of life might have been still more appalling. No trustworthy information was obtainable the day following the accident as to the direct cause of the explosion. The most intelligent theory advanced, however, is to the effect that fulminating caps were carelessly loaded with the dynamite cartridges, and that in some inexplicable way one or more of these caps was ignited causing the explosion.

At Hoge's Point the road makes a sharp bend and a flagman is stationed there, where he can watch the track in both directions. The air was filled with smoke, dust and flying debris of all kinds. The powder had in some way ignited, and the car, engine, tender, passengers, and even the roadbed, were blown several hundred feet in all directions. The flash of the explosion was plainly seen from the station at Tarrytown, about one mile distant, and Officer Smith who was on duty there, John A. Lant, editor; David Whalen, a brakeman, and hundreds of others rushed to the spot. They found the roadbed blown out for a distance of twenty or twenty-five feet, rails twisted into all sorts of shapes, the car almost completely annihilated, and the bodies of the victims strewn all about.

Five minutes after the explosion the smoke cleared away. At least it had all blown out over the river. A few minutes later the workmen who had been cutting stone by the track three-quarters of a mile north came running up. They had recovered from the shock. People from the Tarrytown station came too. They saw dead or wounded men lying everywhere. Fifty feet south of the spot where the explosion had occurred was a pile of human bodies tangled together, legs and arms and heads sticking out of the pile here and there. Nearby was another pile, but not so large as the other. Two bodies floated in the river 100 feet from the shore. The steam escape valve on the top of the engine held one body impaled, the arms and head hanging on one side and the legs on the other. The ground was covered with bits of clothing and pieces of flesh.

Two telegraph poles had been knocked over into Hoe's Pond, and the wires dangled in the water. The fence which had separated the company's property from the pond was missing for 200 feet. The river and the pond were full of floating debris.

The Cranbury Press, New Jersey, **1891-05-22**

Hartford, CT, May 21, 1892, Aetna Pyrotechnic Fire

The Aetna Pyrotechnic Works in this city were destroyed by an explosion of Greek fire this afternoon at 4:30 o'clock. The disaster caused the instant death of five persons, three of whom were women. By the time the Fire Department was able to reach the scene the ruins were in flames.

In half an hour bodies were rescued. It will not be possible to ascertain the real cause of the disaster. The business of manufacturing the Greek fire has been carried on here for eight years, and the works have been located in a thickly-settled part of the city. The Coroner will make a thorough investigation.

The scene of the disaster was nearly identical with that of the great car-workers explosion in 1854, by which fifty persons were killed and wounded. By direction to the scene and immense crowd of people.

The New York Times, New York, NY, **22 Mar 1892**

West Berkeley, CA, July 9, 1892, Powder Works Explosion

An explosion occurred at the Giant & Gunson powder works near West Berkeley shortly after 9 o'clock this morning, destroying those works and much property for several miles around.

Five shocks were felt in this city within a few minutes, the last four being of terrific force, shaking buildings, cracking a number of walls and breaking plate glass in buildings eight blocks from the water front. It is now believed that 104 were killed, including three white men, nearly all the men employed in the works being Chinese. The explosion set the adjoining buildings on fire and owing to the danger of additional explosions no one ventured near the works to stop the progress of the flames. The Giant Powder works are located at Point Isabel and Highland, near Stege, and comprise five buildings and three magazines. Of these buildings the acid and nitro glycerine works are known to have been destroyed, and at least one magazine blown up. Other buildings not immediately destroyed took fire and it is not believed that anything will be saved.

The loss on property will be heavy and it is also believed that there is a great loss of life.

From the top of the hill just above the works the scene beggars description. On the western slope the scattered timbers of the giant powder house blaze furiously, while little yellow streams running down to the bay show where the contents of the acid tanks had emptied themselves. All mixing and packing houses of the black powder department are in ashes and also the sulfur mill, which buildings were located east of the scene of the explosion and on the opposite side of the hill. The damage in the black powder works alone will reach $75,000.

The first explosion caused the giant powder magazine to blow up. The shock traveled in an easterly direction, and the black powder mills lay directly in the path, and burning brands were heaped upon the already wrecked buildings. The cause of the explosion is said to have been the upsetting of a bottle of acid in the office, which set fire to that building. Two injured Chinamen were taken out of the ruins at 12:30 and a number who escaped with burns and bruises are huddled nearby, but are unable to give any account of the explosion, as they do not speak English. The damage is estimated to be at least half a million dollars.

The number of injured has not been definitely ascertained. Later dispatches from the scene of the explosion show that early reports are greatly exaggerated so far as loss of life was concerned. Considering the terrific force of the explosion the loss of life is very small. It is believed it was confined to the three white men. Some Chinamen may have been killed, but it is not known positively.

The Fresno Bee, California, **1892-07-10**

San Francisco, CA, May 22, 1895, Nitroglycerine Factory Explosion

A terrific report and concussion which was distinctly felt all through the city and towns around the bay for a distance of 40 miles yesterday, was at first believed to have been caused by an earthquake, but proved to be an explosion in the nitro-glycerine and mixing houses of the California Powder Works across the bay. The crew in the glycerine house, four in number, and the foreman of the mixing room were killed, as were also nine Chinese working in the latter department. The explosion occurred in the nitro-glycerine house and was probably caused by a Chinese dropping a can of the explosive. The cause cannot be definitely ascertained, however, as all connected with the building are dead.

There were 200 Chinese in that adjacent mixing room, and at the sound of the explosion all ran. The force of the explosion was tremendous. Huge pieces of wood were thrown into the bay, a distance of half a mile, and nitro-glycerin tanks were hurled a distance of 500 yards. Hands, legs and other parts of the mutilated remains of the dead were scattered along the road or a mile. The nitro-glycerin house first went up, and the mixing storehouse and gun cotton in the premises followed.

The nitro-glycerine house of which not a vestige now remains, was a 8-story frame structure, 1200 feet by 50 feet. It contained 8,000 pounds of nitroglycerin and 2,000 pounds of Hercules powder. A remarkable feature of the explosion is that although the storehouse containing 1,000 pounds of Hercules powder is completely wrecked its contents are intact. In all 100,000 pounds of explosives went up with a roar and with a sheet of flame.

Fort Wayne News, Fort Wayne, IN, **23 May 1895**

Murray, KY, February 24, 1897, Dynamite Explosion

At the town of Murray, southeast of here, seven people were instantly killed in a gravel pit this morning by the explosion of fifty pounds of dynamite. The victims were seated around a fire in the pit playing a game of cards. One of them emptied some sawdust containing a dynamite cap on the blaze, which exploded, igniting the box of dynamite. One person was sitting on the box of the deadly explosive, and his body was torn literally to pieces. All of the other bodies were horribly mangled. About eight others standing nearby were badly injured.

Daily Gazette Xenia, Ohio, **1897-02-24**

Cygnet, OH, September 8, 1897, Nitroglycerin Explosion

A terrible explosion of nitro glycerin occurred here at 3 p.m., which resulted in the death of six people so far as known. The explosion occurred at Grant well, located at the rear of the National Supply company's office building, in the village limits. This well had just been shot by the shooter for the Ohio and Indiana Torpedo company. The well was a gasser and when the 120 quarts of nitroglycerine let down into the well exploded, the gas ignited and with a terrific roar the flames shot high above the derrick. As soon as the drillers saw the flames several climbed into the derrick to shut off the gas, but they had hardly gotten there when there was a terrific explosion.

The burning gas had started the remaining glycerin in the empty cans standing in a wagon near the derrick. In another wagon nearby were some cans containing another 120 quarts of the stuff, and this was started by the force of the first explosion. The second was blended with the first in a mighty roar, and the town and surrounding country for miles trembled from the shock.

The National Supply company's building was completely demolished, and nothing remains but a big hole where the wagons stood. There is not a whole pane of glass in any window in the town and every house and store was shaken to its foundation. There was awful excitement over the affair, and all the remaining population of the town rushed to the spot. The identity of the men who were in the derrick and who were killed cannot be learned now, owing to the excitement.

The damage to the Ohio Oil company will amount to $3,000. Eight buildings are a total wreck and many others damaged. The town has a population of about 1,200. Many bystanders were wounded.

Newark Daily Advocate, Ohio, **1897-09-08**

Historical Incidents 1900s

Portland, OR, June 27, 1911, Union Oil Company Fire

An alarm came in from E. Salmon and Water Street shortly after 7:45 a.m. An oil pump at the Union Oil distributing plant had thrown a spark, igniting gas accumulated in its motor pit (Figure 2.139). As he got into his

Figure 2.139 Union Oil Fire. (Courtesy Portland Oregon Fire Department.)

automobile, Chief Campbell knew the fire would be hot. One of the first at the scene, he began directing arriving engine companies. By 8:30, every fire company in the city was on the line, an incredible jumble of men, machines and horses, slipping in inches of water as they tried to position themselves.

Realizing that the fire was out of control and their only hope for controlling it would come with an interior attack, Campbell borrowed a turnout coat from one of his men and disappeared into the building. At 8:39, there was an ominous rumble from the basement as accumulated gases approached their flashpoint. Bodies were hurtled across the street, tank heads flew 200 feet in the air, the north wall was tossed across the street and the roof fell back to the ground. Campbell was last seen silhouetted against the flames, holding up his arms to brace against the falling roof.

By 10:15 a.m. when the fire was brought under control, word had passed from engine company to engine company that Chief Campbell had gone into the building before the explosion and had not come out. Rescue efforts began, and his body was found huddled in his borrowed turnout coat with the letters "F.D." still visible on the buttons.

A Profound Loss for Portland

The passing of David Campbell signaled the end of an era. Campbell had successfully straddled the cusp between the old and the new in terms of manpower organization and technology.

After Campbell's untimely death at the age of 47, the public came out in droves to mourn the hero they fondly knew as "Our Dave". Over 150,000 citizens crowded into downtown Portland streets for Campbell's funeral, which to this day is the largest number of people that have ever gathered for a similar occasion in Portland's history.

Fireman Who Made the Supreme Sacrifice:

Chief David Campbell (Figure 2.140), Portland, Oregon Fire Department

Figure 2.140 Fire Chief David Campbell. (Courtesy Portland Oregon Fire Department.)

Waukegan, IL, November 27, 1912, Corn Products Plant Blast

Through investigation of the explosion which wrecked the starch house of the Corn Products Refining company at Waukegan, Ill., resulting in the death of eleven known persons, the supposed killing of four others and the injury of twenty-five men, women and children, was ordered by Coroner Taylor of Lake county. Three workmen, injured in the explosion, died in the hospital. Seven more were said to be near death. After the coroner's jury viewed the scene of the disaster the inquest was adjourned until after the funerals of the victims.

The coroner arrived at Waukegan on an early train and the inquiry started immediately. The injured were reported in such serious condition that the hospital authorities sent a call to Milwaukee for additional doctors and nurses to care for them. Officials of the company also ordered an investigation. Throughout the night rescuers worked with the aid of lanterns among the ruins in search of the bodies believed to be buried beneath the tons of brick, stone and timber. Wealthy and prominent citizens of Waukegan aided the police and firemen in the work, but when daylight arrived no additional victims had been found. At least four bodies, however, are believed to be in the ruins.

"The accident began with fire," said Charles Ebert, superintendent of the Corn Products company. "The first flash of flame set fire to the powdered starch in the air and the powder exploded with all the force of dynamite. I have been through several disasters of the same sort. They all start that way. They are unavoidable." The bodies of two men were catapulted out of windows sixty feet to a nearby cemetery. Several of the dead had been pinned down by wreckage and were roasted to death in the fire. One victim was pushing a wheelbarrow fifty feet away. The explosion lifted a wall and flung it upon him.

The Marshfield Times, Wisconsin, **1912-12-18**

Baltimore, MD, March 8, 1913 Harbor Dynamite Explosion

Three hundred tons of dynamite being loaded in the British tramp steamer Alum Chine in the lower harbor, off Fort Carroll, exploded about 10:30 o'clock this morning, instantly killing from 40 to 50 men, wounding and maiming three-score more, some of whom may die, and dealing destruction to half a million dollars worth of property (Figure 2.141). The Alum Chine and a loading scow alongside here were completely annihilated; the tug Atlantic, which twice went to the rescue of imperiled seamen, was set on fire and later sunk; the United States collier Jason, just completed and ready for trial, was raked to her deck and her armor riddled, and buildings in Baltimore and cities and towns many miles away were rocked by the force of the terrific explosion. The steamer was loading dynamite for the Panama Canal.

The cause of the disaster is unknown tonight, but Federal authorities have instituted a thorough investigation to place the blame. Excited

Figure 2.141 Three hundred tons of dynamite being loaded in the British tramp steamer Alum Chine in the lower harbor, off Fort Carroll, exploded about 10:30 o'clock, March 8, 1913. (*Baltimore Sun* Files.)

survivors told conflicting stories, some insisting that a stevedore caused the explosion by jamming a pike into a case of dynamite. This is denied by eyewitnesses, who declare that smoke was seen pouring from the Alum Chine's hold several minutes before the explosion occurred.

The bodies of 20 dead have been brought to morgues in this city, and 60 injured are in the hospitals. The estimates of the dead include 30 stevedores and checkers employed in transferring dynamite from a barge to the Alum Chine, which was bound for Panama, 8 members of the crew of the Alum Chine, 6 men on the collier Jason, and the captain and seven members of the crew of the tug Atlantic. Many bodies, it is believed, never will be recovered from the icy waters.

Of the injured a score are frightfully maimed, their skulls fractured, arms and legs torn off, and their bodies terribly lacerated. At least fifteen are expected to die. Pieces of machinery and portions of the hull of the boat, weighing tons, were shot hundreds of feet into the air. Pieces of iron and steel 3 feet long and weighing 50 pounds were found at points on the Anne Arundel and Baltimore county shores, 3 and 4 miles from the scene of the explosion. Seamen in scores of small craft and Captain, of the Atlantic, saw smoke coming from the dynamite-laden Alum Chine a few minutes before the explosion, and the launch Jerome was alongside the ship, rescuing many members of the crew and rushing them to safety. The witnesses say that when the explosion came the steamer jumped from the water as if a torpedo had struck her from below, and then fell to fragments, in which were mingled the, torn bodies of the dead.

The transport company's scow had been tightly tied to the big steamer, and the concussion from the ship's hold blew up the tons of dynamite still aboard that barge in an echoing explosion that increased the carnage and destruction. Several small craft in the harbor are believed to have been blown to pieces. The tug Atlantic, which had twice rushed in to the rescue before the crash, was moving away when it was caught in the rain of charred wood and red-hot steel that fell in a shower for a quarter of a mile around, killing some of the crew outright and setting fire to the vessel.

Coming up the river at the time was the big Britannia, and she rushed to the aid of the Atlantic, picking up many of the wounded who had leaped into the water. Some of them are said to have perished before the Britannia could get to them. The Britannia ran a line to the burning rig, and started up the river with her, but the vessel sand before the Lazaretto lighthouse was reached. Great was the relief when it was ascertained the cutter and crew were safe. "The Guthrie steamed away, going about a mile to wait for an incoming vessel, and it was while we were moving around that the captain and others noticed fire on the steamer.

This appeared to be coming from the forecastle. Knowing the danger, Captain of the cutter, ordered every window opened so the explosion expected would not break the glass. He then steamed toward the burning vessel, but when within three-quarters of a mile the ship exploded. The Guthrie got a severe shaking, trembling like a reed in a storm." "Following the explosion we saw a huge inky cloud which ascended fully 200 feet and almost covered the harbor. When this cleared away there was naught in sight. Where the steamer had been loading, the vessel, scow, and cars had disappeared. We hurried to the scene, but saw only wreckage."

"Our firemen, who were shoveling coal, got the full force of the explosion in their faces. Many of them were burned. I expected to have 150 men in the crew by tonight, and about 100 men were on board." Immediately after the first shock of the accident the Captain ordered his men to make a close inspection of the entire ship. They found many holes in her side. One hole was big enough for a man to put his head through. This hole is eight feet above the water line. Other holes are near the water line. "I can picture it now. It seemed like a great column of fire 50 feet high and 20 feet across, topped by another column of black smoke 200 or more feet higher, came up from the sea, completely enveloping the ship. It was several minutes before the smoke cleared away and the sea became calm, but when it did there was no sign of either the ship or the barge that was alongside of it. They both seemed to have disappeared completely, and not a sign of life was visible."

How the fire started in the coal bunkers of the Alum Chine is so far a mystery. Some have ascribed it to spontaneous combustion, while one of the rescued declares he saw a fellow stevedore stick a bale hook into a box of dynamite. This explanation appears improbable. Had such an incident occurred there would have been an immediate explosion. Property Loss $600,000.

The Washington Post, District of Columbia, **1913-03-08**

Buffalo, NY, June 25, 1913, Grain Elevator Explosion

According to the estimate made this afternoon by the police, twenty-six of the men who were employed in the mill when it was blown up by an explosion yesterday afternoon are still unaccounted for. It is now known that five men were hurled to instant death, eleven are missing, all believed to be dead, and sixty one burned and bruised victims are in local hospitals, as a result of the explosion yesterday afternoon in the grain elevator of the Husted Mill Company's plant, caused by spontaneous combustion. The loss of property is over half a million.

Firemen at work in the ruins this morning found the torso of a workman lying under burned freight cars. The head, arms and legs of the man had been eaten away by the flames. There was nothing by which identification could be made. To the horror of the firemen and spectators, the bodies of two men could be seen on a disjointed upper floor slowly being consumed by the flames. Streams of water were turned in this direction to prevent total consumption by fire but every now and then the flames would roar anew. The heat even this morning from the fire was so intense that rescue of the bodies was impossible at 9 o'clock.

Fifty were taken to various hospitals, some in a serious condition from burns and broken legs and arms. At the scene of the accident a priest administered the last rites to 26 men who were believed at the time to be mortally hurt. That many of the injured cannot survive their wounds and that the ultimate death list may reach fifteen is an estimate reached by a canvass of the widely scattered hospitals where the injured were taken.

The explosion was caused by the puffing of dust accumulations in the feed house and was of frightful force, tearing out the north wall of the wooden structure and breaking windows for a quarter of a mile around. The engineer of a switch engine on the New York, Chicago and St. Louis railroad, was blown from his cab and received internal injuries from which he died later.

Several windows in cars of a passing Nickel Plate passenger train were broken by the explosion but none of the passengers were injured. The train was brought to a standstill and many of the injured elevator men were placed in the baggage car and rushed to the station where ambulances awaited them. Every ambulance in the city was in commission. A twelve year old unidentified boy running to the fire was struck by an automobile and instantly killed.

Another body was blown fifty feet by the explosion and was found under a box car nearby badly burned and mangled. The elevator company employed the men and according to the company all have been accounted for except four. Employees who escaped were equally positive that many of their fellow workmen were cut off by the flames and perished. Ten men were rescued from the roof of an adjoining feed house by firemen. They had leaped from the elevator and several suffered broken legs.

Olean Times, New York, **1913-06-25**

Cleveland, OK, March 9, 1916, Nitroglycerine Explosion

Six hundred quarts of nitro-glycerine exploded near Cleveland, Okla., late today, killing one man and almost causing a panic in Tulsa, thirty-five miles distant, when the concussion shattered windows and rocked frame buildings. Many persons here ran terrified into the streets, fearing an earthquake. The driver, who was hauling the explosive to the Cleveland oil field, was blown to atoms by the accident. The cause of the explosion has not been learned, although it is thought that the nitro-glycerine was detonated by the jar of a rough stretch of road.

Titusville Herald, Pennsylvania, **1916-03-09**

Philadelphia, PA, September 8, 1917, Frankford Arsenal Explosions

Five buildings at the Frankford government arsenal here are in ruins, two workers are dead and 30, including several women are injured, following a series of five explosions at 2 o'clock this morning. Two employees have not been accounted for. Col. Montgomery, commandant of the arsenal, would make no statement regarding the cause of the explosion.

Among the workers, the first report was that a six-inch shell into which powder was being pressed in the high explosives building R. A. house No. 7 exploded. A later explanation was that a workman dropped a three-inch shrapnel shell which exploded when the percussion cap struck the floor, throwing shrapnel in all directions. The two bodies have not been identified. One was in a condition which will prevent identification.

A badly burned unidentifiable body was also removed from the wreckage. The buildings destroyed were two R. A., or artillery assembling houses, numbers 7 and 6, and three I. X. or dry houses. The R. A. houses are houses in which high explosives are handled and the loading of big shells is part of the work carried on. The I. X. houses are those in which powder, gun cotton and fuses are dried. All the buildings were frame except R. A. house No. 7, a new brick structure.

According to unofficial information, the series of blasts was started in the new brick high explosives building, known as P. H. No. 7, when powder being pressed into a six-inch shell exploded. In the high explosives or P. H. building, high explosives are handled in a number of ways and one of the purposes of the building is the loading of big shells.

At the bursting of the big shell, the building broke into flames which shot quickly to P. H. building No. 5, igniting the explosives therein and sending that building up in flames. The flames leaped quickly to three dry houses, called sometimes detonator houses in which powder fuses and gun-cotton are dried. All the destroyed buildings were about 25 yards apart.

Work in the arsenal was ordered suspended at once by Colonel Montgomery, the commandant, but at 6:45 he rescinded this order and ordered the workers admitted as usual.

Confusion both in the plant and in the surrounding neighborhood followed the series of blasts. As the men and women employed in the arsenal rushed from the buildings their relatives living nearby ran to the scene in thousands, frantic with fear.

Heavy rain which fell during the night is believed to have prevented worse damage than resulted. The soldiers on duty said without the rain they would have been unable to keep the flames away from the large magazine, where thousands of pounds of powder are stored.

Oscar Wistner, Jr., who with his sister Clare was at work filling shells in the loading room in plant R. A. 2, told this story of the accident:

"Suddenly there was a blinding flash and a terrific explosion which threw us to the ground. When the shock was over there was a mad stampede for exits. I was trampled upon by several men and I know many others were in a like predicament." "I got out of the crush and searched for my sister. Flames were shooting about in many directions. Girls screamed, while others were moaning from injuries. I found my sister covered by a mass of wreckage and dragged her, unconscious, to safety." According to Wistner, shrapnel flew in all directions through the crowded workroom. He says it is miraculous that so many escaped unwounded.

Affecting scenes occurred when terrified mothers, wives and sisters, huddled in crowds about the arsenal gates, and soaked to the skin by the rain which fell throughout the night, greeted their loved ones, whom they had feared were buried beneath the blazing wreckage.
Chester Times, Pennsylvania, **1917-09-08**

Jersey City, NJ August 17, 1915, NJ Oil Plant Explosion

One man was burned to death, two were so badly scorched that they will not live until morning and five others were injured yesterday afternoon by the explosion of a big tank of gasoline in the laboratory of the Eagle Oil Works, Cavan Point, Jersey City. The cause of the explosion is not known. An employee was tightening a valve on a 64-gallon drum of gasoline and the others were working about the laboratory when it occurred. It is believed that a spark generated by the friction of the valve screw ignited the gasoline. The fire ran along the floor and ignited a number of oil barrels standing in one corner. These were extinguished by other employs, who hurried to the rescue. A hurry call was sent for ambulances from the Jersey City and St. Francis hospitals. It is estimated that about $1,000 worth of damage was done to the laboratory. MR. HAY, the chief chemist, although badly burned, refused to go home. He remained to superintend the repair work, which was started immediately. The Eagle Oil Company is a subsidiary of the Standard Oil Company.
New York Tribune, New York, **1915-08-17**

Ardmore, OK, September 27, 1915, Gasoline Explosion

The death list in the gasoline explosion here was raised to fifty when the bodies of four unidentified workers were found in the wreckage of a building on Main street about noon. The death list is growing hourly as mangled and charred corpses are dragged form ruined buildings. There are over 200 injured in local hospitals and private homes. Five of these died during the night and it is reported that many more can live but a few hours. Thirty-five bodies were recovered last night. Six more were taken from a wrecked pool hall near the scene of the explosion early today, where seven other bodies were found last night.

Witnesses were found today who declared there were two big explosions. This was borne out by the fact that there was a large quantity of dynamite stored in the demolished freight shed. A second detonation following immediately after the first is believed to have resulted from the ignition of the dynamite. It developed today that there was but one workman employed in repairing the tank car in the Santa Fe railroad yards, when a spark from his hammer ignited the gasoline and resulted in an explosion as terrific as an equal quantity of dynamite. The workman was Ira Woods, employed by a refinery company. Bits of his body were found many yards from the scene of the explosion. The workman who was supposed to have been helping Woods left the car for a few minutes and escaped injury. A railroad car containing casing gas exploded, killing 45 people and destroying much of downtown. Fort Wayne News, Fort Wayne, IN, 28 Sept. 1915.

Syracuse, NY, July 5, 1918, Split Rock Explosion

Forty-nine men are known to be dead. Parts of sixteen more charred bodies are reported to have been found in the debris of the plant. The number of dead was said this afternoon to be at least sixty-five. At least fifty men were missing this afternoon and it is feared that the death list may reach seventy-five or even one hundred. Half a hundred injured are in various Syracuse hospitals. Some of these will die. This with $1,000,000 property damage is the toll of the explosions which wrecked half of the great munitions plant of the Semet-Solvay company at Split Rock at 9:30 o'clock last night. Forty-eight mangled, or burned bodies are stretched side by side upon the white-tiled floors of the County Morgue. All available employees of the company to-day were digging in the ruins. They found blackened bones and parts of bodies, heads, arms and legs. Hour by hour during the night and morning the death list grew. Hour by hour, ambulances and the morgue car brought in more bodies. Many are missing. Relatives of men employed at the plant sought them at the morgue. They stormed the gates to get in, all through the night and through the morning after.

As body after body was brought to the morgue in Montgomery street the crowd of watchers pressed forward trying to follow ambulances and dead wagons through the gates. Guards held them back. The gates were closed and locked. Up to late in the afternoon there were about twenty-nine of the bodies at the morgue unidentified. The explosion, which shook the entire city and villages nearby occurred at 9:30. A fire starting at 8:40 led to the explosion. The main blast, followed by minor blasts, could be heard for miles in every direction. A statement given out by H. H. S. Handy, president of the company says that the fire started in T. N. T. plant

Figure 2.142 A statement given out by H. H. S. Handy, president of the company says that the Split Rock fire started in T. N. T. plant No. 1 from an undetermined cause. (Courtesy Syracuse.com.)

No. 1 from an undetermined cause (Figure 2.142). At first the fire did not appear alarming but it spread to T.N.T. In the process of manufacture and several explosions followed.

A statement by Burton J. Hall indicates that the explosion was in his nitrofier in Plant No. 1. where 2,200 pounds of T.N.T. Were in the first stage of manufacture. Men in Plant No. 1 heard a warning to "get out of here" at 8:40 o'clock. They rushed from the building but later went back and manned several lines of hose to play on the fire. It was while they were fighting the fire that the explosion occurred. Flying debris struck many of them, killing or injuring. In an instant, scores of men were knocked down or torn to pieces by the explosion. Some were blown through walls. They were blinded temporarily or made unconscious. After the main explosion flames rose to a great height and smoke was carried for twenty miles. Some hurried to telephone to the city for doctors and ambulances. Patrolmen rushed to the rescue but many of them had been killed.

Nitrofier Caused the Explosion Says Victim

Burton J. Hall Relates Remarkable Experience Before and After Horrible Blast at Chemical Plant. The probable cause of the explosion is given by Burton J. Hall, No. 1012 Montgomery street, who had a remarkable escape from death. Hall thinks his nitrofier in Plant No. 1 blew up. This contained 2,200 pound of toluol. "There were but two nitrofiers in this building," Hall said. "I was working 2,200 pounds in my nitrofier," said Hall. "It was in the first stage. There was but one other nitrofier in the plant and the toluol there had not reached the stage where it would explode." Hall's story is a graphic description preceding and after the explosion. He runs a nitrofier. "At 8:40," he said, "while I was at my place taking the temperature somebody shouted: 'Get out here" Black smoke was then coming from the opening from downstairs.

There were four in the room, and myself. We were on the top floor of No. 1 and made a rush out. This brought us on the level of the hill. We ran to the road, then to the police station and with about twenty others grabbed hose lines. Potter and I had one hose. Potter told me to go in the building and turn on the water." "I went in and turned it on about three-quarters, but the smoke was so dense I had to come out. Others were playing water on the fire, but we are we went up on the hill where a freight car stood." "This car was loaded with acid and with about twenty or thirty others we pushed it down the track for we thought it might explode." "There were five or six around me when the explosion came. I was thrown flat on my face and was unconscious. Later I couldn't see anything." "But when I came to I found that my friends were lying all about but that only my back and shoulders were hurt. I reached down to pick up my hat and discovered that my coat and my trousers had been blown off." "Instead of my own hat I picked up a patrolman's hat. Putting my hands over my head I ran to the hill, then I could begin to hear things. And it looked like a war picture with the smoke and flames and everything."

There was the usual crowd of morbid sensation seekers. These the police held firmly in check and they got no further than the front door. Those who could prove that they had relatives who worked at the Rock were permitted, in groups of four or five, to view the bodies.

Of the forty-seven bodies which had been brought to the morgue at noon only sixteen had been identified. Coroner Crane at this time declared that it is doubtful if the others will ever be known, so frightfully are they disfigured. One of these, the wearer of a big ruby ring, may be made known before night through this bit of jewelry.

The search for lost relatives steeled many women for the ordeal of viewing the bodies. There were one or two cases of collapse, but that was all. The failure of some men to return home after the accident brought many people to the morgue. Some of them failed to notify their families of their safety and this gave rise to frantic endeavors on the part of women to get in touch with Split Rock, to discover if the men were still working. "My husband didn't come home last night. May I see the bodies? Is he here?" was the general question.

Relatives of those who had been absolutely identified sought to take the bodies away at once, but this was denied by Coroner Crane. "This afternoon, "he said" we may let some of them go out, but until we are positive that we can't identify any more they should not be removed. The removal of any bodies at this time would give erroneous ideas to some people that there had been a mistake and that perhaps after all their own husbands or brothers had been brought here and removed."

The Syracuse Herald, New York, **1918-07-03**

Vallejo, CA, November 8, 1918, Mare Island Navy Yard Explosion

Six persons are dead as the result of a terrific explosion which destroyed the black powder storehouse of the magazine at the Mare Island navy yard early Monday. Thirty-one injured persons have been taken to the navy yard hospital. The list of wounded includes non-commissioned officers, enlisted men of the United States navy and civilian laborers. The explosion occurred at 7:45 a.m. Eye witnesses said a column of black smoke arose 300 feet. Shells, parts of steel plates, huge splinters of wood and debris were hurled in all directions.

Vallejo is separated from Mare Island by a narrow channel, which accounts for the damage in this city. A few minutes after the explosion all the naval officers at Mare Island and about 1,000 marines and other recruits were hurried to the scene to fight the fire, which was soon under control. The first men reported injured were three on the Southern Pacific ferryboat El Capitan. The ferryboat was in the stream about two miles from the scene of the explosion, and was damaged considerably, all the windows and doors of the boat being blown out. At the Southern Pacific freight sheds in Vallejo, about two miles from the scene of the explosion, the doors of the buildings were torn from their hinges.

Santa Rosa, about forty miles northwest of Mare Island, reported that the explosion was felt there. Residents thought it was an earthquake. San Francisco Agents of the Department of Justice were sent to Mare Island by United States District Attorney John W. Preston to co-operate with the naval authorities in determining the cause of Monday's fatal powder explosion.

Fairplay Flume, Colorado, **1918-11-08**

Big Heart, OK, January 26, 1919, Nitroglycerine Explosion

Eight people were killed and more than a score severely injured when a wagon carrying nitroglycerine belonging to the Eastern Torpedo Company exploded in the heart of the residence district here today.

The wagon driver and a passenger also on the wagon were blown to atoms. A residence in front of which the explosion occurred was leveled to the ground. Occupants were perhaps fatally wounded and their three-year-old baby boy was killed. Seven other houses in the vicinity were wrecked. The explosion broke every window in the town and shook the ground for hundreds of yards around. All telegraph and telephone communication was destroyed. Big Heart has only one doctor. He had a corps of workers attending to the dead and wounded. Pawhuska, Ok, sent physicians and rescue workers in motor cars to the scene. Not all the bodies of the dead and wounded were recovered from the ruins of the houses

and casualties may exceed first figures. Only two quarts of nitroglycerin were in the wagon. The cause of the explosion is unknown.
Lima Daily News, Lima, OH, **26 Jan 1919**

Cedar Rapids, IA, May 1919, Douglas Starch Co. Explosion

That the management of the Douglas Starch works is lacking in its efforts to complete an intelligent check on the missing and the dead is apparent from its stubborn intention to keep the public as much in the dark as possible regarding the catastrophe. Naturally, the officials could not be expected to make a detailed statement, theoretically or otherwise, concerning the tragedy in its plant, but every attempt to obtain some sort of information regarding the identity of the men known to have been in the plant at the time of the explosion, as compared to the list of unaccounted for, was flatly opposed by the management and its representatives.

Shortly after noon today eleven bodies were in the undertaking establishments of the city. Eight of them had been identified either positively or partly. The list of identified was changed during the morning. In the houses near the plant where the windows were blown out, plaster knocked from the walls and ceilings and other damage done, many little children were injured by the flying glass. They are in need of medical attention. Already the work of relief has been started. There are about eighty men in the R.O.T.C. force from Coe that is doing guard duty around the plant. A strict guard line is maintained. Passes are necessary both to go in and go out of the roped area.

Yesterday the largest exclusive starch factory in the world was humming with activity. Today the two huge stacks and the wet starch building at the west end of the plant are all that is left of the great plant. White smoke rose from the smoldering ruins that were once the dry starch building. It was there that the big explosion happened. In the large building just south flames are still to be seen. But the fire has been under control for many hours and the firemen are standing by still playing streams of water keeping the fire from increasing and gradually putting it out.

There is not a single house near the plant that does not have a broken window. Most of them do not have a single one left. Besides smashing windows and the blast rocked the ground, jerking plaster to the floors, tore shingles from the roofs, and caved in the sides of the buildings. Many were cut by the flying glass or bruised by the falling plaster. It is impossible to obtain even a partial estimate of the number hurt in this way. It is estimated that the total loss to the plant may total $3,000,000.

The Red Cross got on the job immediately after the blast. Nurses went to the scene of the disaster and civilian aid was begun at once. The guard house for the Coe men who walked post on the downtown streets during the night was in the Hotel Montrose. These guards were to stop petty thieving in the windows that had been laid open by the explosion. Most of the merchants cleared their broken windows of all the merchandise so that there was little temptation for sneak thieves.

The blast was felt all over the city. Soon afterward the flames caught fire. Several men were taken from the ruins and rushed to the hospitals. Others were heard shouting or moaning beneath the debris but these soon stopped. In some cases it was the creeping fire that stopped the cries. It was announced shortly afternoon that the company would have a public statement to make at 3 p.m.

South side business houses suffered wide damage from broken windows. Not a single store front with large windows escaped on either side of the river. The Citizens Savings Bank on the west side and the Iowa State on the east side suffered the heaviest loss in that section of the city. All of the lower floor huge windows on the north side of the Iowa State were blown in. Riverside park was strewn with wreckage carried there by the force of the explosion.

Among the heaviest losses suffered in the business district was that of the Overland-Doty Automobile company. It occupies a new building at Third avenue and Eighth street. The entire front is of plate glass and not one of the windows is left.

The ruins were still smoking this morning but the fire had been under control for some time (Figure 2.143). Practically all of the bodies taken from the debris had a hand raised as if the men had intuitively raised a hand to ward off something when the explosion came and before they were killed, either burned to death or crushed beneath the falling stones,

Figure 2.143 The Douglas Starch Company ruins were still smoking this morning but the fire had been under control for some time. (*Cedar Rapids Gazette* Archives.)

bricks and timber. Trade in the stores was practically at a standstill this forenoon. Clerks were busy assisting in clearing up glass of discussing the event. However it was not expected that business would be retarded although some noticeable decline might be expected from the shock to the community and the unsettled conditions.

The firemen were hampered at the start by the crowd, who in their desire to help got in the way. Some who tried to assist the firemen seized the hose and with several on each pulled in opposite directions. With the arrival of the special police and the R.O.T.C., the confusion was abated and the work of rescue was organized. One of the injured men who was taken out of the ruins was buried under several large beams, iron rods and brick. He remained conscious although he was badly injured and directed the firemen in their work of rescue.

Other bodies taken from the spot were badly burned and in one case it was necessary to pick up the bones piece meal. Another body was taken out with the head missing. The firemen had difficulty in making their way about, the water thrown on the ruins resulted in the starch and dust to become exceedingly slippery and as a result the fire boys are hobbling around with stiff limbs. Thirty five missing, may or may not be in ruins, six identified dead, five unidentified dead, ten at St. Luke's hospital, eight at Mercy hospital.

Three and four streams of water are being played on the ruins all the time by the fireman. The wet wood and the steam rising make a white cloud. Some of the stock on the table house, the building just south of the dry starch building, is still burning. But there is no danger of the flames spreading.

The Evening Gazette Cedar Rapids, Iowa, **1919-05-24**

Whiting, IN, July 10, 1921, Standard Oil Explosion

Five men were burned to death and twenty-seven were injured when the fire started today in three batteries of high pressure stills in the Standard Oil Company's refinery at Whiting, Indiana. An explosion followed the fire. The property loss was said to be heavy. Among the dead were Fire Marshal Johnson and his assistant, Joe Taylor, of the Standard Oil Company's force. The other victims were laborers and were so badly burned that identification will be difficult. Seven persons were reported missing. It was not until noon that the fire was extinguished and it was then that the search for the bodies was made. It is believed there are some bodies in the ruins. The flames broke out just as the day force was leaving and the night force coming to work. The damage to the refinery was not estimated. Some officials said it was incalculable and that the cost of repairs would be huge.

Logansport Morning Press, Logansport, IN, **5 Jul 1921**

Wellington, KS, June 7, 1922, Tank Car Fire

While flames fed furiously on a stream of gasoline from a damaged tank car, in heat so intense that he could scarcely breathe, Switch Engineer Eldridge Paxton battled the flames with steam from the blow-off cock from his boiler until the flames were overcome here early this morning. The flames spurted all about him. Paxton and his crew had gone to the rescue of two switchmen hurt in the collision which damaged the tank car and caused the gasoline to spurt forth.

Gasoline spurted over other nearby tank cars and the flames followed it. Although the crow said they expected to hear a terrific explosion momentarily, the stream subdued the flames before much damage was done. The gasoline was ignited after the collision by the passing of another engine.

Tulsa World, Tulsa, OK, **7 Jun 1922**

New York, NY, July 18, 1922, Chemical Explosion

One fireman was killed, many policemen, firemen and spectators injured, and scores of others had remarkable escapes this morning in a fire and series of explosions in the six-story warehouse of the Manufacturers Transit Company, at No. 10 Jane Street, near Greenwich Avenue. The fire was discovered shortly after 8 o'clock. Assistant District Attorney Morgan Jones said he had learned that the building contained large quantities of magnesium, sulphur and potash. A combination of these makes the flashlight powder used by photographers. There was also a quantity of celluloid and a shipment of German toys belonging to the New York Merchandise Company, and there was much rubber.

Frederick Francis, Treasurer of the company that operates the warehouse, first refused to say what chemicals were stored there, saying he would have to see his lawyer first. The owner of the building is Edgar Bluxton, who lives at the San Remo Hotel. A rough estimate of the property loss is $1,000,000, including the value of the building as about $300,000, print paper of the Star Publishing Company at about $300,000, and paper belonging to the Tribune Company, at $30,000.

Mayor Hylan arrived early at the scene and wading through puddles of water, in some spots half way to his knees, crossed over to the west side of Fourth Street and plunged into the heavy pall of smoke in search of "Smokey Joe" Martin. The Assistant Chief came out of the building and shaking hands with the Mayor, said: "Mr. Mayor, this is the toughest and worst fires I've ever witnessed in all my career in the department." Going closer to the fire than anyone except the firemen had been allowed, Mayor Hylan made a survey and then announced he would order a full investigation as soon as he returned to City Hall.

Fire Commissioner Thomas J. Drehan declared there was "something to be called to the attention of the District Attorney" regarding the fire. An early report that an engine company had been lost entirely to an explosion that tore off part of the roof proved to be erroneous but the real story of the incident is remarkable. The men concerned in it were six members of Rescue Squad No. 1, under Lieut. Kilbride. They were playing a hose down into the flames. Kilbride and Fireman Charles Rogencamp were at the nozzle. When the explosion came those two and their four helpers were blown back across the roof to the edge of the coping, where they lost their hose and scrambled to their feet.

A part of the coping dropped to the roof of a five-story tenement at No. 16 Jane Street, and there was another explosion at once which wrecked the two upper floors - from which, fortunately, all tenants had been removed a few minutes before. The members of the Rescue Squad found themselves stunned by this second explosion, but recovered and made their way down the fire-escape.

Acting Chief "Smoky Joe" Martin, who was blown through a door and slightly injured, said the building apparently contained quantities of magnesium and phosphorus, which added to the difficulties. Water poured on burning phosphorus accomplished nothing.

Fireman Who Made the Supreme Sacrifice

Lt. John. J. Schoppmeyer, NY Fire Department, Engine No. 13

Injured Firemen

Alexander, Phillip J.; Engine Company No. 14; cuts.
Brown, Harry, fireman; Hook and Ladder Truck No. 5.
Burke, Anthony E.; Hook and Ladder No. 3; lacerations to left leg.
Butts, William, fireman, Engine Company No. 33; cuts.
Calamari, Michael, Fire Patrol No. 2; both hands badly burned.
Coleman, P. J.; Policeman, Charles Street Station; bruises.
Corkery, Edward, fireman, Engine No. 33; cut and bruised.
Donlin, John, Hook and Ladder No. 3; blown through a roof scuttle by explosion.
Doughty, Thomas, Hook and Ladder No. 3; both legs crushed.
Havilane, Rudolph, Engine Company No. 72; eyes affected by smoke and chemicals.
Kahn, Morris, Engine Company No. 33; lacerations.
Lewis, Joshua, Engine Company No. 3; cuts.
Lynt, E. D., fireman, Engine No. 13; cut and bruised.
McCaffrey, Michael, Lieutenant Fire Department.
McConville, Hook and Ladder No. 5.
Martin, Joseph, Acting Chief, Fire Department; cut and bruised.

Mullaly, Edward, Captain of Engine Company No. 74; lacerations of
   left foot.
O'Brien, James, Engine Company No. 21; overcome by smoke.
O'Connor, Capt., Fire Department Engine No. 18; cuts.
Reilly, Patrick, fireman, Truck No. 5.
Reynolds, Peter, fireman, Engine No. 14; overcome by smoke.
Rosenkamp, Charles, Rescue Company No. 1; injured hand.
Rotunno, Salvatore, Engine Company No. 24; knee and hand injured.
Williams, Stephen, fireman, Engine 33; cut and bruised.
Yonholtz, Charles, fireman, Hook and Ladder Truck No. 5.

The Evening World, New York, NY, **1922-07-18**

Altoona, KS, February 12, 1924, Nitroglycerine Explosion

A man believed to be Harry Percival of Independence, Kan., was killed
instantly today when a motor truck, heavily loaded with nitroglycerine,
was blown up three miles north of Altoona. He is said to have been an
employee{sic} of the American Supply Company of Independence. The
shock of the explosion was felt for miles around Altoona, reports said,
and it was thought to be an earthquake.
   Reno Evening Gazette, Reno, NV, **12 Feb 1924**

Pittsburg, PA, August 20, 1924, Gasoline Explosion

An explosion of gasoline last evening in the garage of the People's Natural
Gas Co., Forbes Street, took a toll of eight lives, three of them boys, caused
probable fatal injury to seven persons and less serious injury to seven others. A part of the People's Natural Gas Co. building was wrecked and
adjoining buildings were shaken by the blast. The probable damage was
estimated at $10,000.

The explosion occurred when a 1,500-gallon tank of gasoline was being
drained into the storage tank at the garage. Four automobiles in the garage
and six others standing in the street were destroyed by fire which followed
the blast. Flames shot through windows and doors of the building from
the truck which had hauled the gasoline tank into the garage. Officials
of the company declared they could assign no cause for the blast. Scenes
of wildest confusion followed until police reserves arrived and established
fire lines. Windows within a radius of eight blocks were shattered.

The majority of those seriously injured were standing near the garage.
The clothing was burned from some of them and they were hurled thirty
feet by the concussion. Automobiles were commandeered by the police
to remove the injured to hospitals. About twenty-five persons were in
the building at the time, the majority of them being girls. Three young

women were among those injured. They were hurled across the room by the blast. The others were hurled to the street from the third floor, where they were working.

Pittston Gazette, Pennsylvania, **1924-08-30**

Langtry, TX, March 5, 1925, Quarry Explosion

Langtry, Tex. Eight men were killed and four were seriously injured in a premature explosion in a Southern Pacific Railway Company quarry here. A score of men working at the quarry escaped injury, but could not tell how the accident occurred. It is presumed that one of the charges of dynamite was imperfect.

A train carrying physicians, nurses and medical supplies arrived from Del Rico to render aid. The bodies of some of the victims were blown to bits. Others were hurled more than 200 yards while some were buried beneath heavy masses of rock and earth as the side of the hill toppled over on them.

Palo Alto Reporter, Emmetsburg, Iowa, **1925-03-05**

Clinton, IL February 21, 1925, Acid Explosion

Clinton, Feb. 21. Harry Sutton, thirteen year old son of Mr. and Mrs. Roy Sutton, was terribly mangled and in all probability fatally injured after 11 o'clock Saturday forenoon, and two companions of about his age were slightly hurt by flying glass as the result of a terrific explosion of acids contained in two bottles the lads found while playing in the Argo pasture east of the Illinois Central tracks. Young Sutton was still alive in Jarman hospital here at twelve o'clock, though it was believed he could not survive long.

When found by the three boys, the bottles of liquid were picked up by young Sutton, who crashed them together. Instantly a violent explosion occurred, mangling young Sutton horribly and blowing a hole in the ground in a depth of three feet. Particles of the broken bottles were hurled against the other two boys, cutting them slightly in several places.

Later it was determined that one of the bottles contained nitric acid and the other sulphuric acid, the contact of which caused the heavy explosion. It is believed the bottles were left where they were found by [ineligible] for future use, perhaps in Clinton.

Harry Sutton's father, who is a brakeman in the employ of the Illinois Central, was on the St. Louis end of his run when the tragedy occurred, but is rushing toward this city as fast as possible. He will arrive here at 5:30. Mrs. Sutton, the mother, is being sought. Neighbors say that she is out of the city, but her whereabouts is unknown to them. A message to her will have to await the arrival of her husband.

Decatur Daily Review, Decatur, IL, **21 Feb 1925**

Pensacola, FL, January 2, 1926, Newport Tar & Turpentine Co. Explosion & Fire

With the known dead numbering twelve, ten injured in hospitals and five workmen missing, firemen continued tonight to battle flames which resulted today when a fire retort at the Newport Tar and Turpentine company exploded, with expectation that the bodies of the missing would be recovered from the ruins.

Fire officials estimated the damage to the property at $200,000. The blast which came a little more than an hour after the plant opened today, demolished one unit of the large plant in which 22 men were working. Flames spread in the resinous premises and soon engulfed other units. Some of those who were injured were believed to have been standing near the unit which exploded and were caught before they could flee to safety.

The cause of the explosion was undetermined tonight. Many of those killed were badly mutilated and some of the bodies are still in the burning unit. It will probably be impossible to remove them tonight, officials said.

Firemen reported they could see three bodies behind a barrier of flame in the until where the explosion occurred. It was believed several of those who are missing may have been trapped by smaller explosions in some of the other burning units.

Nevada State Journal, Reno, NV, **3 Jan 1926**

Sparrows Point, MD, November 20, 1926, Oil Tanker Explosion

Painstaking search of the cooling ruins of the Norwegian tanker Mantilla, rent apart by an explosion in its hold while in dry dock at Sparrows Point, early today had accounted for only 8 of 16 men believed killed. A check made by Captain Nils Danielsen indicated that the death toll included 12 of the Mantilla's crew of 35. Four unidentified bodies were apparently those of seamen. The other bodies recovered were believed those of employees of a ship cleaning company. Of these, the body of Amiel Petersen of Baltimore was identified.

Of the two score injured receiving hospital treatment, several had slight chances of recovery, physicians said. Many injured went to their homes.

Accidental ignition of gas generated from oil that remained in a fuel tank when the Mantilla went into dry-dock was believed the cause of the blast which ripped the vessel's steel plates like tissue. Flames licked from the fissures, scaring workmen from the docks and dry-dock scaffolding. Rescue workers, who penetrated billows of smoke which blanketed the dry-dock after the explosion found disabled men frantically attempting to escape from advancing streams of burning oil. Many were taken out with their clothing afire. Thorough search of the interior of the vessel was impossible until several hours after the explosion.

At that time no trace of bodies could be found. Officials believed the missing men may have been hurled into the water, or consumed in the flames. Lars Larson, second mate of the Mantilla, who was aboard when the vessel blew up, gave a graphic account of the disaster from his bed in the marine hospital. He was seriously burned.

"Of a sudden I heard a terrific crash," he said, "and saw a vivid flame that seemed to shoot out from the side of the ship. It seemed as if the whole side of the ship was coming down on me." "During this time I heard continuous explosions. Not so loud as the first, but with each successive bang there was a crash and part of the ship seemed to give way."

The Mantilla was owned by W. Wilhelmson of Ponsberg, Norway. She was being reconditioned for service of the Mexican Petroleum Co., on the Tampico-Baltimore run.

The explosion was the most serious harbor disaster in 13 years.

North Adams Transcript, Massachusetts, **1926-11-20**

Alton, IL, June 15, 1926, Refinery Explosion

Four men were killed and five injured in an explosion and fire at the refinery of the Standard Oil Company of Indiana at Wood River, near here, today. According to a statement by company officials, the explosion, the cause of which is unknown, occurred in the liquid asphalt mixing tank. A pitch still nearby was ignited where were employed the men killed or injured. The men killed and injured, it was said at the office and of the Standard Oil Company, were employed on a pipe line close by the pitch stills. The bodies of the men killed were burned so badly that identification was difficult.

Investigation by company officials following the tragedy revealed that gas apparently had escaped from one of the petroleum stills and that extraordinary atmospheric conditions had permitted it to collect close to the ground.

The gas was ignited by fire beneath one of the stills, and the resulting explosion and swirl of flames burned almost beyond recognition the four men working on the still. One of them, William H. Koehne, 25, of Wood River, a helper on one of the pitch stills, was responsible for detecting escaping gas, C. B. Manndeck, superintendent of the plant said, and with Kohne's death the chance for discovering the immediate cause of the explosion was lost.

"Outside gas explosions are very rare," Manndeck said. "Unusual atmospheric conditions must have prevailed to permit saturation of the air by the gas." Koehne was leaving his still when the explosion occurred, on his way to shut off the steam pump which forces the liquid into the still.

Joplin Globe, Missouri, **1926-06-15**

Turner, ID, January 19, 1927, Gas Explosion in Mormon Hall

Six persons were killed and nearly a score injured, 12 seriously, as a result of an explosion of acetylene gas, which wrecked a Mormon chapel and recreation hall here last night while a basketball game was in progress. About 200 persons had assembled in the one story frame building for a game between the Turner and Central, Idaho, teams. Shortly after play started, the lighting system failed, and witnesses said, someone lighted a match. A terrific explosion followed, wrecking the rear wall.

As the players and spectators started for the only remaining exit at the front of the building, a portion of the ceiling fell, hurling timbers and plaster into the crowd. Before the hall could be cleared the front wall collapsed. Most of the dead and many of the injured were found near this exit. The body of James McCann, the janitor, was found in the basement where he had apparently gone to inspect the lighting system tanks. The bodies of his two sons and his brother, Brigham McCann, were found near the exit. Two more bodies were identified as those of Elmer Anderson and Iral Lowe, both members of the Central team. Mrs. Lowe and Mrs. Andeson, who had accompanied their husbands to the game, were injured. Some of the injured said they were literally blown from the building.

Iowa City Press-Citizen, Iowa, **1927-01-27**

San Pedro, CA, February 15, 1927, High School Classroom Blast

In a series of explosions followed by a fire that turned the laboratories of the San Pedro high school into an inferno of blazing chemicals, 15 students and two teachers were painfully burned or injured today. Miss Capitola Nunn, a teacher, of Los Angeles, and Loire Gentry, a student, were the most seriously injured and are under the care of physicians. Others were treated for first degree burns and sent to their homes.

The students were removed to the San Pedro General Hospital, where they are suffering from burns about the arms and face. The blast occurred when an experiment being conducted in the laboratory exploded, shooting flaming chemicals about the room. The blaze swept through the laboratory and destroyed the adjoining classrooms before firemen brought it under control.

The 1200 students, including the injured made their way to safety without aid.

No estimate of the damage could be made, but the entire section of the building occupied by the laboratory was said to have been badly damaged by the blast and subsequent flames.

Oakland Tribune, California, **1927-02-15**

Chicago, IL, March 11, 1927, Chemical Company Explosion

The ranks of engine companies numbers three and six were depleted late today when an explosion killed one fireman outright and seriously injured ten other, two probably fatally. They had controlled a fire in the plant of the Daigger Chemical Company on West Kinzie Street, just outside the downtown district, when a terrific explosion occurred and was followed by several of lesser degree. All the other injured firemen suffered serious body burns when their clothing caught fire.

The firemen killed and injured were trapped in the basement. Fumes filled the air and the men, bruised shocked and burned, were brought out only after comrades had donned gas masks.

Five firemen collapsed at a street door they sought to escape from the gas-filled basement and were dragged to safety by bystanders.

Max Woldenberg, president and several other officials of the chemical company, were taken to the police station for questioning. Twenty-five employees of the chemical company fled the building just before the explosion occurred. More than 200 workers in nearby buildings fled to the street when they were rocked by the explosion. Carbide of sufficient quantity to damage the buildings in the entire block was found in the basement of the two-story brick plant, Samuel Kugelman, battalion chief said, explaining that the chemical explodes upon contact with fire or water.

M. J. Corrigan, another battalion chief, in charge of an investigation, said the chemical company had been under surveillance for some time in connection with reports that chemicals were stored in the basement.

Firemen Who Made the Supreme Sacrifice

Edward Hirsekorn, Truck 6, Died on the Scene
Tomas Bender, Truck 3, died the next day, March 12, at Henrotin Hospital.

Injured Fireman
Patrick Kelly inhaled flames and was burned about the hands, face and body.
Independent, Helena, Montana, **1927-03-12**

Akron, OH, April 30, 1928, Benzene Explosion

Four boys, three of them brothers, were killed and a fifth was seriously injured when a steel drum of benzene exploded in a shed at the rear of the Goodyear Tire & Rubber Company plant here Sunday.

Those killed were:

Charles Carter, 14.
Albert Carter, 10.

Leeman Carter, 5.
Frederick Wadtly, JR., 13.

Leslie Bush, 14, was badly burned although physicians believe he will recover.

The drum exploded when one of the boys lighted a match fumes apparently were leaking from the container and as the match ignited there was a terrific explosion. The five boys were deluged with blazing benzene.

One of the lads was killed instantly. Screams of the others brought nearby residents to the scene. They wrapped the boys in blankets and rushed them to hospitals where they died, one by one. Albert Carter was said to have struck the match. Leeman Carter staggered to a point across the street from the blast where he collapsed. His brothers managed to crawl some distance before they too collapsed, unconscious.

The Wadtly boy was killed by a piece of flying steel. The shed had been used as a storage house for fire hose. The drum had been standing in the shed for several months, persons living in the vicinity said. The public relations department of the Goodyear company disclaimed ownership. Herbert Maxon, public relations director, began an immediate inquiry which, he said he hoped would prove ownership. Police likewise began an investigation. Fragments of the drum were blown 100 feet. Only the lid remained. The shed was partially destroyed.

The scene immediately following the blast was poignantly tragic. The fathers of the Wadtly boy—drawn to the scene by the detonation—hastened to call an ambulance. He did not know his son was among the victims. The lad's sister stopped her father. "I believe Frederick is one of those hurt papa," she said. Wadtly returned. He identified his boy by a mouth organ found in his pocket. A spirited marble game preceded the explosion. A call "to come to supper" broke up the game and as the lads prepared to disband, Albert stepped to the benzene tank and struck the match. The body of the Wadtly boy was hurled into a pile of stacked tiling.

Parents of the boys were grief-stricken and tragic scenes were enacted as the burned and bruised forms were wrapped in the blankets. A triple funeral service was held for the Carter brothers.

Sandusky Star Journal, Ohio, **1928-04-30**

Kokomo, IN, May 12, 1928, Explosion in Laundry

An explosion occurred in the steam pressing department of the Fridlin Laundry here today. Four women were killed and seven others were reported seriously injured. It was feared some of the injured might die. The laundry is located directly in the center of the city. Many windows in buildings nearby were shattered. The concussion was felt for several miles. So great was the shock that tracks of the Union Traction Co. running in the center of the street in front of the laundry, were torn up for several feet.

The entire laundry building was wrecked. Workmen in all sections of the building were thrown from their feet and it was believed that many were hurt. The entire city was thrown into confusion. A number of persons were injured, several seriously. It was impossible to get a complete list of the injured, due to the confusion after the blast. Company officials declared the steam boilers could not have exploded. They said pressure was below normal and they were at a loss to explain what had caused the explosion.

Star Journal Sandusky, Ohio, **1928-05-12**

Elizabeth, NJ, February 19, 1930, Oil Refinery Explosion

The ranks of 60 injured workmen, who were seared with exploding naphtha at a Standard Oil refinery late yesterday, were depleted by death today until the list of fatalities had reached 10. Among the 50 bandage swathed laborers in three hospitals here, were 20 whose chances of recovery were doubtful. It was feared that several of these would be permanently blinded if they survived the scorching blast which spurted thru part of the bay way. The explosion occurred yesterday afternoon in the alcohol plant of the refinery. J. Raymond Carringer, general manager, said the blast was caused by the breaking of a gas line.

Survivors said the pipe which broke was a high pressure naphtha line and that fumes from it were ignited by portable forges used by workmen constructing a new building near the one story building housing the alcohol plant.

The night shift of 1,100 men employed on the 100 acre reservation occupied by the plant in Linden, a suburb, had started work at 3 o'clock. They had been on duty 55 minutes when there was a terrific explosion followed immediately by two lesser blasts. Long tongues of blue flame from alcohol fumes shot from the windows and doors of the building and from gaping holes torn in the roof and walls. About 20 feet from the alcohol plant a force of masons, carpenters and laborers were working on the new building. The force of the explosion threw them from scaffolds while the flames set their clothing afire.

Workmen employed in the alcohol plant rushed out, their clothing aflame and many blinded by the fire. They left four of their fellow workmen dead in the plant. Their bodies were recovered after the fire which followed the explosion had been extinguished. The four others died after being taken to hospitals. At the hospitals cots were set up in corridors to care for the injured. Every doctor and nurse in Elizabeth was called upon.

William Slaven, a mechanic, who came out of the alcohol plant unscathed said:

"It seemed as if flames were shooting out in long tongues all around me. Many of the tongues of flames acted freakishly as they followed the path of the alcohol fumes. The flames were on all sides of me. One sheet of flame was only 10 feet from me. It left me untouched, but I saw it strike some other men."

James Caperinicchio, a laborer employed on the new building, said he was wheeling a barrow of mortar when the explosion occurred. He said although he was 150 feet from the alcohol plant, he was hurled to the ground by the blast and found himself afire. "I leaped up," he said, "and everybody around me looked as if they were dead. They were all on the ground. Then they began getting up. Most of them were on fire. They started running. There is a heavy wire fence around the building just there.

Most of the men seemed unable to see. They couldn't find the gate to get out and they became panicky, beating at the fence and screaming like mad." Besides burns suffered from the flames, many of the workmen on the new building suffered fractures and other injuries in falling from the scaffolds.

Mason City, *Globe-Gazette*, Iowa, **1930-02-19**

Norfolk, VA, January 8, 1931, Oil Barge Explosion

Norfolk spent last night fighting a three-million-dollar waterfront fire. Started by an explosion on an oil barge at the Buxton line piers, the fire rode a strong wind along 200 feet of wharves, destroyed a hotel, swept through the wholesale district and leaped across Main street to threaten the retail business center.

The explosion, the cause of which was sought by firemen today, occurred late yesterday, and it was after midnight before the combined fire fighting forces of the entire Hampton roads area, aided by 1,300 sailors and marines, brought the fire under control. A driving rain that began just before midnight checked the spread of the flames. The Victoria hotel, a 200-room structure, was destroyed. Miss Carrie Ambrose, switchboard operator, remained at her post until all guests and employees had escaped. The American Peanut corporation plant was burned at an estimated loss of one million dollars. Seven city blocks were swept.

During the night 30 persons, including firemen, reported to hospitals for treatment and 16 of them has injuries sufficiently serious to cause them to be kept as patients. A large number of others received treatment at ambulances stationed near the fire zone.

A force of 1,300 blue jackets and marines from battleships Arizona and Utah and from the naval base was called out to maintain order.

Equipment and firemen from Newport News, Portsmouth, South Norfolk, Suffolk, the naval base and Norfolk naval yards, were called to the assistance of Norfolk fire fighters. In addition to the business houses burned the T. J. Hooper, a 200-foot tug belonging to the Eastern Transport company, an unidentified tug and a smaller barge, together with 200 feet of wharf, were damaged.

The Charleston Daily Mail, Charleston, WV, January 8, 1931

Long Beach, CA, March 1, 1931, Gas Explosion

Three oil workers were dead Sunday victims of an oil well gas explosion in the Signal Hill district. The dead were E. H. and H. H. Hanley, brothers, and H. S. Clark.

Dallas Morning News, Dallas, TX, **2 Mar 1931**

Kilgore, TX, April 18, 1931, Oil Tank Explosions

The death toll of a two-day series of freakish oil field fires here was raised to eight today when C. E. Upchurch died in a Tyler hospital from burns received last night in a blaze which also proved fatal to J. W. Smith. The same fire seriously burned L. H. Gray and R. C. Holcomb. They were being treated at hospitals, but their conditions were not believed to be critical.

The fire started when a workman struck a match near the tank battery of the Upchurch & Allen No. 2 Brightwell well. The well had come in during the day as a big producer. It was being cased and there was heavy gas pressure. A 60,000 gallon tank of oil was ignited by the gas flame and destroyed. Upchurch owned an interest in the well, and Gray is vice president of the Sabine Pipeline Company. Mrs. Kate Dobson, her stepson and two children were burned to death in another fire yesterday when gas and oil close to the tent in "Tent City" near here in which they were sleeping caught fire.

The first of the series of fires burned to death Joe Lamb and Milton H. Petteway on Thursday. This fire was caused by a spark from an automobile exhaust igniting a gas pocket near a battery of oil tanks.

Bradford Evening Star, Pennsylvania, **1931-04-18**

Marcus Hook, PA, February 5, 1932, Tanker Explosion

While investigators scanned the hulk of the oil tanker Bidwell, which blew up at Marcus Hook early yesterday, the death toll today was placed at eighteen with one missing and a woman and three men battling for their lives in Chester Hospital. The boat was shattered by four terrific blasts while lying in dock at Marcus Hook. Questioning of injured survivors failed today to explain the tragedy.

Most of the crew were asleep or on shore leave when the blasts tore open one side of the ship. The death list may mount today. Olaf Rasmussen, 32, of Forest City, Ia., is believed dying of his burns and injuries at the Chester Hospital, Chester. Others in the hospital, less seriously injured, are Mrs. Viola Rivers, of Jamaica, N.Y., wife of the captain; Lero McMahon, 27, of Fostoria, Tex., and William Major, of Chester. Twelve dead were counted at the Chester morgue, and others were given up for lost and their names added to the gruesome list.

After a systematic search of the hold of the steamship Bidwell for four missing members of the tank cleaning gang proved futile, it is now believed the bodies of the men were burled into the river and drowned. Searching parties have been engaged by the Sun Oil Company to watch and grapple for the bodies. It is said by experienced river men that, owing to the cold weather, it may be some time before the bodies come to the surface.

As searchers finally were able to board the ship which for hours had been a raging inferno, bodies of men charred beyond recognition were found. Still more bodies, blown into shapeless masses, were located aboard the ship of death. More victims, injured beyond recovery, died at hospitals. Sun Oil Company officials are at a loss to account for the cause of the blasts. In an effort to solve the mystery surrounding the origin of the first blast, which ignited the ship and led to the other explosions, a minute examination of the charred and warped debris of the steel vessel is now in progress.

As the toll kept mounting hourly, officials still were unable to determine what mischance had set off the first devastating explosion at 12:20 a.m. That first blast, which turned what had been an ordinary boat into a scene of horror, was followed by three other blasts within twenty-five minutes. The countryside within a radius of twenty miles had been shaken and rocked. Thousands of sleeping persons were awakened and terrified.

It "might have been" a spark which set off a gas pocket in the hold of the ship while the tanks were being cleaned of the residue of crude oil—which had been brought from Texas and unloaded Tuesday—so that gasoline could be put into them, some investigators said.

The charred body of Captain Rivers was found in his quarters. He had gone back to salvage his logbook and personal records and paid with his life for his devotion to the duties of the sea—to save the log if possible.

Several other members of the crew missed death or injury by having gone ashore for the night. Included among these are John Petterson, John Ball, a cook, and Terrence O'Connor, 21, of 748 East Westmoreland street, Philadelphia. O'Connor missed death through his devotion to his widowed mother. Her husband died only recently and she was supporting a family of six with the help of Terrence. He had gone home for the evening to take $5 to her to "help out a bit." That saved his life. Ball, a World War veteran in describing the horror, said it "scared him" and reminded him of bombardments in France.

And so the stories of survivors went. Almost to a man those who had come through the thing alive had done their level best to save a fellow worker. Into the record of the oil company, however, were inscribed a saga of heroism on the part of the members of the crew of the ill-fated ship many of whom braved death to rescue comrades who had been disabled by the first blasts. Another risked his life to save the life of Mrs. Viola Rivers, wife of the captain, the only woman aboard.

The first revelation of the heroism of the crew came when the charred body of Captain Joseph Rivers, skipper of the tanker, was extricated from the blackened hull of the wrecked ship. True to the traditions of the sea, he had stood by his ship to the last, perishing in his attempts to save his men.

Further disclosures of the bravery of the crew came last night when Coroner J. E. Scheehle revealed how Edward Cartain, 29, of Buckman Village, near here, had rescued Mrs. Rivers from certain death. She had been tossed in the Delaware River and was being drawn towards the flaming ship when Cartain jumped from shore, where he had reached safety, and finally managed to bring her ashore, although he was almost exhausted.

"His was the most startling example of bravery," said Coroner Scheehle. "The man, who apparently was on the dock, rushed to the ship when the flames started to rage, jumped into the water and swam to the hole which had been torn in the side of the ship." "Mrs. Rivers was floating nearby unconscious, and was being carried slowly towards the opening. Undoubtedly she would have been drawn into the inferno in the hold. She wore a life preserver and bobbed in a sea of burning oil."

"In another minute she would have been in the flames, but Cartain, who certainly ought to be recommended for a medal, reached her in time and then swam heroically with her to the dock." "Then with an unheard of modesty he disappeared," Coroner Scheehle said.

Many of the bodies of the dead were charred almost beyond recognition.

Chester Times, Pennsylvania, **1932-02-05**

Berlin, NH, January 23, 1934, Oxygen Cylinder Explosion

Frank McKay, local truck man employed by W. H. MacArthur of Berlin, narrowly escaped death Monday when one of more than a dozen oxygen cylinders, each containing 200 cubic feet of the gas, exploded while he was leading them onto a truck, blowing the body of the truck to bits.

The accident occurred at the local Boston and Maine freight depot. The explosion could be heard all over the city. It sent the other tanks soaring through the air over a radius of many yards but no other cylinder was believed to have exploded, although one or two remained unaccounted for (for) several hours later.

McKay said he did not remember what happened. The last thing he knew, he said, he was in the act of loading one of the cylinders on the truck. He "came to" some minutes after the explosion to find himself walking around near the wreckage of the truck. He bore no outward signs of injury but was taken to a hospital for examination and held overnight.

Portsmouth Herald, Portsmouth, NH, **23 Jan 1934**

Norman, OK, June 5, 1934, Dynamite Explosion

The mangled bodies of seven men lay in funeral homes here today, victims of an accidental explosion of dynamite intended for use in seismograph explorations for petroleum deposits. As relatives sped toward Norman to assure themselves of the identification of their dead, undertakers and their assistants continued piecing together the shattered bodies of the members of the Petty geophysics engineering company crew who met horrible death on a rural roadside eight miles southeast of here late yesterday.

Officials investigating the accident were at a loss to explain its cause, although some attributed it possibly to static electricity, others to a conceivable collision between a "magazine" truck carrying dynamite, and a company water truck also found at the scene. Dynamite is one of the requisites for carrying on the geophysical explorations for oil, in which the men were engaged. The charges of the explosive are detonated over the area to be surveyed, and delicate instruments are used to record the sound wave echoes of the blasts as they bounce back from rock structures far underground. By timing the interval between explosion and echo, depths to the underground structures may be ascertained and an accurate knowledge obtained of the subsurface topography.

One of the few persons near the fatal blast was Wendel Crawford, another member of the party who was making instrument observations while mounted on a truck about 50 yards away from the seven. Evidently under great physical strain, Crawford described his impressions.

"My vision was completely obstructed," he said, explaining that there was a "partition" which separated him from the rest of the crew. "When I saw what had happened, I turned my truck toward Norman and reported to company headquarters and called ambulances."

"That is all that I can say."

Denton Record-Chronicle, Texas, **1934-06-05**

Asheville, NC, December 25, 1938, Fireworks Plant Blast

An explosion in a downtown fireworks store tonight killed at least three persons and injured an undetermined number. An hour after the explosion firemen, backing their way through the wreckage, had brought out three bodies, and expressed fear several others had been killed. At that time the firemen had not penetrated to the basement of the building, which caved in.

It was said a number of persons were in the store. Christmas is observed throughout this region with fireworks displays. Automobiles parked on both sides of the street were damaged by flame and heat. One vehicle directly in front of the building was demolished.

Bud Shaney, professional baseball pitcher, said the fire probably was caused by an open charcoal pot around which employees were gathered. Shaney said six to ten persons were in the store when he left. The explosion occurred when he was about 25 feet from the entrance, he said.

Lawrence Stepp, 15, who was in the store at the time of the explosion, said he was unable to find his mother, Mrs. Maude Stepp, who was standing beside him at the time. Young Stepp said he was knocked into the middle of the street. The store, he said, was in flames and firecrackers were shooting across the street. The boy was deafened in one ear. He said he was sure the blast was caused by the fire in the charcoal pot.

Galveston Daily News, Texas, **1936-12-26**

Owensboro, KY, November 12, 1939, Glenmore Distillery Fire

A wild fire raged out of control Saturday night at the Glenmore Distillery, one of the largest in the country. Company officials had small hope of saving the plant from complete destruction. An unofficial canvas of insurance men placed the loss already at $2,225,000.

Flames roared through three warehouses, the bottling plant and the company office and was enveloping the distillery building. The plant covers four city blocks and employs 700 persons. Fire of undetermined cause tonight swept through a section of the Glenmore distillery, one of the largest in the country, razing five buildings and destroying thousands of barrels of whisky. More than 50 firemen from four towns battled the flames for five hours.

James Pendleton, managing editor of the Owensboro Messenger said unofficial estimates of insurance men placed the loss at between $2,000,000 and $3,000,000. One fireman was injured when a barrel of whisky exploded knocking him to the floor. He was removed to a hospital.

Dallas Morning News, Dallas, TX, **13 Nov 1938**

Camden, NJ, July 30, 1940, Camden Paint Factory Explosion

At least two persons were killed and eight others were missing tonight in a devastating fire that raged for nine hours in the huge paint factory of the R. M. Hollingshead Company. A state of emergency was declared in the city by Mayor George Brunner after the fire, starting with a series of explosions, spread through the plant and to dwellings several blocks away, driving hundreds of families from their homes.

As the fire was brought under control by firemen at about 10 o'clock tonight, the eight missing, four men and four women, were believed dead in the vast smoldering ruins which firemen were still drenching with water. The first known casualty, Raymond Carter, 38 years old, of Collingswood, N. J., an employee of the factory, was blown from the factory by one of the explosions and died in a hospital tonight.

Beginning shortly after 1 p.m. with a series of terrific explosions that were felt for miles around, the flames swept through the paint factory, destroyed or badly damaged 125 near-by frame dwellings, caused injuries to at least 205 persons, including fifty firemen, and resulted in property damage tentatively estimated at $2,000,000, Five hundred National Guardsmen of the 157th Field Artillery began patrolling tonight an area of sixteen blocks from which approximately 600 persons had been evacuated.

They were on guard to prevent efforts at pillaging the abandoned homes and business places. One man was arrested for looting and was almost lynched before the troops took charge of him. At the height of the conflagration late this afternoon, when eight fire companies and 100 policemen from Philadelphia were aiding the entire Camden Police and Fire Departments and numerous volunteers from near-by New Jersey communities,

Mayor Brunner declared the state of emergency to conserve water for use in fighting the flames. Appealing to large water users in the city to shut down until the fire could be brought under control. Mayor Brunner assigned fifty city employees to ride through the streets in the northern part of the city shouting, "Don't use water," to all who would listen.

Fireman Who Made the Supreme Sacrifice

William Merrican, 49, a fireman of the Camden Fire Department who collapsed in the intense heat of the blaze and died at a hospital of a heart attack.

Camden County News, **July 30, 1940**

Toppenish, WA, November 29, 1940, Dynamite Warehouse Explosion

A death dealing explosion followed by fire ripped a large warehouse asunder at Toppenish, Wash., today, killed at least seven persons and left an untold number of other buried beneath the ruins. The blast was believed caused by gas which collected in the warehouse basement of by dynamite which was stored there. About 12 persons were injured, six were taken to hospitals in serious condition. The blast reduced the two-story, 250-foot warehouse to rubbish. Destroyed in the warehouse were a merchandise store, a real estate office, barber shop, beauty shop, produce company office, and a cafe. The explosion apparently occurred in the center of the building. The roof collapsed and the four concrete walls crumbled.

Independent, Helena, Montana, **1940-11-29**

Du Quoin, IL, February 15, 1941, Liquid Oxygen Plant Explosion

At least five separate investigations were under way today in the explosion that claimed the lives of seven employees of the United Electric Coal Companies' liquid oxygen plant at the Fidelity mine near this city yesterday. But Fred Huff, superintendent of the mine, said the cause of the blast was as much a mystery as it was Friday. Besides the company's investigation, others were being conducted by Robert M. Medill, state director of mines and minerals, by Coroner W. E. Gladson and by insurance officials.

In addition it was known that agents of the Federal Bureau of Investigation had entered the case but secrecy enveloped their activities. All seven victims were instantly killed and there were no injured survivors to aid the investigators. Some of the men were unloading lamp black cartridges from a freight car on a siding a few feet from the plant. Others were working in the plant where the cartridges are soaked with liquid oxygen, converting them into explosives used in blasting coal and rock.

Neither the lamp black nor the liquid oxygen is explosive by itself. So far the investigation has not disclosed, Huff said, whether the explosion occurred in the freight car which arrived Wednesday night from Samford, Tex., or in the plant. Both the car and the plant, a two-story frame building, were demolished. Huff said the property damage was "pretty close to $200,000." He disclosed that much of the plant equipment was manufactured in Germany and cannot be replaced at this time because of the war.

Meantime the mine's 275 employees were idle today as survivors of the victims prepared to bury the dead. Separate funerals tentatively are planned for all seven tomorrow. Coroner Gladson empanelled a coroner's jury to investigate the deaths and announced an inquest probably will be held next week.

Edwardsville Intelligencer, Illinois, **1941-02-15**

South Charleston, WV, Nov. 6, 1941, Plant Explosion

A localized explosion at the biggest plant in the great Kanawha Valley's multi-million dollar chemicals industry Thursday killed at least two me, injured five others and started a fire which was still burning long after nightfall. Officials of the Carbide and Carbon Chemicals Corporation plant, which normally employs 4,200 men in manufacturing important defense items, withheld comment about the damage.

Flames prevented entrance to the three-story distillation unit building and it could not be determined whether any other workmen besides the dead and injured had been in the structure.

Unofficial reports placed the customary crew in the unit at between eighteen and thirty men. Some workmen said, however, that only unit operators are on duty at lunch hour and that the blast occurred at 12:32 p.m., just two minutes after lunch period ended. Immediate identification of the victims was impossible, officials said. Only fragments of the bodies were recovered and the identification discs ordinarily worn had been lost in the explosion. Only one of the injured men was reported in serious condition. He was H. R. Fox, 32, who may lose his sight from burns.

Dallas Morning News, Dallas, TX, **7 Nov 1941**

Browntown, WI, December 25, 1941, Silica Plant Explosion

A mysterious explosion, followed by fire, yesterday destroyed the main building of the Browntown Silica company, which had been engaged in supplying silica, used in processing steel, to defense industries. No one was injured. P. W. PALMER, one of the owners, estimated damage at $150,000. FBI agents at Milwaukee were notified.

Charlestown Daily Mail, West Virginia, **1941-12-25**

Versailles, PA, May 2, 1942, Torpedo Plant Blast

The cause of an explosion at the Central Railways Signal Company Plant in nearby Versailles Borough, which killed eight women employees, injured 12 others and wrecked a building in which railroad signal torpedoes were being made, remained a mystery today.

Three FBI agents, state, county, and local officials, who began investigations after the blast blew the roof off the torpedo building and hurtled the workers' bodies against the concrete walls late yesterday, said no evidence as to the cause was found.

The plant, comprising four single-story buildings, makes torpedoes, flares, and other signaling devices used in railroad operation. Investigators said that as the company's output is wholly for private industry, they knew of no reason for sabotage. At Washington, Secretary of Labor Perkins ordered William T. Cameron, chief safety engineer of the Labor Department, to the scene, about 15 miles southeast of Pittsburgh.

The tremendous detonation as the torpedo building blew up was heard for miles. Pieces of its steel roof were thrown several hundred feet.

Windows in buildings blocks away were knocked out. Fire following the blast was quickly extinguished. The company estimated the total loss at $25,000.

Gust Brassler, 50, a railroad brakeman who was the first to enter the blasted building, said: "It was terrifying. Girls were running about screaming, blood gushing from wounds in their legs and arms. We began dragging out all those we could reach." Paul Bodnar, whose sister Helen was among the dead, was standing near the plant. He rushed in to seek her and stumbled over her body. Smoke and flames drove him out. Helen Maravic, 30, an employee who escaped injury, thought the plant was being attacked by enemy planes. "It sounded like a bomb," she related. "I thought we were being attacked just like in the movies. Then everything went to pieces as I rushed out the side entrance." Allegheny County officials asked Edmond P. Thomas and D. S. Kingery, explosive experts of the U.S. Bureau of Mines to conduct an inquiry.

The plant employed 40 persons. The women ranged in age from 19 to 36. Police Chief W. O. Kennedy of Versailles said an explosion at the plant several years ago cost a dozen lives.

Weeping relatives filed through the county morgue to identify their dead, whose bodies were horribly burned and mangled. Identification in some cases was possible only through fragments of broken eye-glasses, a scar incurred years before, or a pair of shoes. Imbedded in the bodies were dozens of small torpedoes, which are made of potash and sulphur and can be exploded only by detonation, officials said.

Open boxes of the potash sulphur mixture, in a section of the plant next to the torpedo room, were still unexploded after the blast and fire. Frank W. Slaney, superintendent of the plant, was in his office adjoining the torpedo room when the explosion occurred. He escaped injury. But his daughter, Elizabeth, 19, working in the torpedo room, was instantly killed.

Injured, taken to McKeesport Hospital told their stories to Coroner William D. McClelland and deputy coroner.

"I groped blindly to find the door," Josephine La Presto, 22, said. "I tripped over someone's body. God only knows whose it was. Smoke, dirt and fire were flying all over the place. About four girls left the building with me." Heavily bandaged Julia Shaffer, 34, also told a tale of horror. "We were trapped. There were men there. We were screaming and begging for them to get us out. Suddenly my hair caught fire, and the men showed us the closest door."

Transportation of the injured from the plant to the hospital was facilitated by a special war emergency ambulance corps which had just been formed, said William A. Hacker, superintendent of the hospital.

Police and firemen from Versailles, McKeesport, Boston, and Glassport, state police from the Elizabeth sub-station and county police from Pittsburgh rushed to the scene.

For hours, Joseph Janus, a Versailles volunteer fireman, worked amid the wreckage aware that his sister-in-law, Susan Kossuth, had been working in the torpedo room. Later he identified her body at the morgue, solely from her hair. K. B. Smith, conductor of a railroad yard crew, said injured women, "burned black and with their clothing torn and smoking, came running out of the building."

Mrs. Mary Rupinski, who lives across the street from the plant, said the explosion frightened her "almost to death." "I thought it was the first air attack of the war here," she said. "I had a daughter and a sister in there, but they're safe, thank God."

Four of the seven injured remaining at the hospital were reported in serious condition. They were John Wall, 69, Versailles; Olympia England, 26, McKeesport; Myrtle Weiner, 32, Boston, Pa.; Julia Shaffer, 34, McKeesport.

Indiana Evening Gazette, Pennsylvania, **1942-05-02**

Smithfield, NC, March 7, 1942, Ammunition Truck Explosion

At least five persons were killed and more than 100 were reported injured near here early Saturday in the delayed explosion of a fire-ridden ammunition laden truck, the detonation of which was heard over a radius of 50 miles in eastern North Carolina. A hotel, a filling station and a tavern were leveled to the ground and windows in Selma, a town about two miles away, were shattered. The cement highway under the truck was ripped wide and deep for a distance of 30 feet.

Dr. E. N. Booker, Johnston County Coroner, said that at least five persons were injured fatally, two of them in an automobile which he said failed to observe highway patrol warnings. The others were killed in the leveling of the Talton Hotel. Dr. Booker said that George Stroup of Gastonia and Cecil Propst of Lawndale, in the automobile, were told that they would proceed around the truck, earlier involved in a collision with an automobile at the roadside community, at their own risk.

"They decided to take the chance of proceeding on their way," the coroner said, "and just as they got almost even with the truck it exploded with a terrible noise which was heard in Rocky Mount, 50 miles away, reported they heard." Two of the killed in the hotel were Mrs. Minnie Lewis, Raleigh, N.C. and Buck Mitchell, 46, Dunn, N.C. The injured were taken to hospitals in Smithfield, Goldsboro and Raleigh.

Dr. Booker said that all buildings within some 300 yards of the explosion were badly damaged or leveled and that half the store windows in Selma were blown out.

Most of the casualties were bystanders who had watched the truck burn for almost two hours before exploding. State Highway Patrolman H. C. Bobbitt said the explosion, heard 25 miles away, occurred at 3 a.m.

The auto-truck collision took place at 1:15 a.m. and fireman summoned from nearby Selma and Smithfield had succeeded in extinguishing the fire in the automobile when they were forced to return to their stations for water.

Bystanders and persons in the nearby Talton Hotel and Gurkins Tavern in the roadside community watched as the flames gained headway again and soon consumed the truck.

"Suddenly the truck exploded and Luke Capps' filling station, about 150 yards away, and behind which I was sitting in my car, was leveled to the ground," Bobbitt said. "My car was demolished and only by the grace of God am I living to tell the details." The truck, Bobbitt said, was owned and operated by Hughes Transportation Corporation of Charleston, S. Car., and driven by Raymond Avery, also of Charleston, who was only slightly injured. It was enroute to Baltimore.

"The automobile caught on fire almost immediately after the collision and sprayed the truck with gasoline," Bobbitt said. "But the truck did not immediately catch fire." "The firemen were making fine headway in controlling the blaze when the water gave out," he continued.

"We then cleared the highway of spectators for a distance of at least 200 yards but several of the buildings destroyed were within the area which we had cleared and some of those injured were in the buildings. The long burning fire, which lit up the heavens for miles around, naturally attracted folk from a wide area and it was difficult to keep them back from the scene."

Mason City, *Globe Gazette*, Iowa, **1942-03-07**

Fontana, CA, December 25, 1942, Kaiser Steel Mills Butane Explosion

Fire caused by exploding butane gas today caused estimated $75,000 damage at the Kaiser steel mill near Fontana but is not expected to delay the start of production scheduled for Dec. 30. The butane gas was being transferred from truck tankers to three 10,000 gallon storage tanks this afternoon when a hose connection snapped. The three workmen on the scene told T.M. Price, superintendent of construction at the mill, that they expected an explosion and fled.

They had barely reached a point of refuge when a series of explosions took place. The blasts were so heavy that wreckage was thrown 500 feet, destroying the three storage tanks, two trucks, an automobile and causing other damage. The butane in the storage tanks was to have been used to start operations at the mill now under construction.

However, Price said that he expects all damage to be repaired in a few days. After the mill gets under way gas generated from the coke ovens will be sufficient to replace the butane The blast could be heard for miles.

Residents of Fontana believe a major disaster had taken place at the plant and rush calls were made to the offices of the Sheriff and Coroner.
Los Angeles Times, Los Angeles, CA, **26 Dec 1942**

Westport, CT, May 3, 1946, Trailer Truck Explosion

State police at Westport barracks said Thursday night a fireman was killed and "ten or twelve" others badly burned when a large trailer truck which had left the road and overturned, caught fire and exploded. At Norwalk hospital a spokesman said that "ten or twelve" men apparently "very badly injured" had been brought to the hospital. No names were available, the hospital reported.

At state police headquarters the desk officer said that a detail had been dispatched to the scene of the accident, which occurred shortly before 11 p.m. on the Boston post road a mile from the center of the town. An officer on duty at the Westport fire headquarters said he had only fragmentary information as the department still was at the scene, but he understood that the injuries occurred when "something in the truck's cargo" exploded.

An explosion which suddenly sprayed blazing liquid on firemen and others as they fought a blaze in an accident damaged truck today left a toll of two dead and eight severely injured.

The truck, reported by police to have contained nylon yarn and rubber cement, exploded on the Boston Post Road here last night. John Lowell, the truck driver, tentatively listed as from Providence, R. I., died early today in the nearby Norwalk Hospital where eight others were reported in grave condition.

Five other Westport firemen and a 16 year old boy were reported to be "very seriously injured" at the hospital where it was said a complete check of the number injured and the extent of their injuries had not been completed. Chief of Police John A. Dolan, an eye witness of the accident, said that the truck, headed toward New York, blew a front tire and swerved into a tree. Almost immediately he said, the flames flashed from the gas tank, enveloping the cab. The driver struggled out and collapsed.

Before the department arrived, Chief DOLAN, said, he heard several muffled explosions from within the truck. When the department truck reached the scene the firemen approached the truck with the hose from the chemical tank just as the door blew out and the men were sprayed with a blazing liquid.

The police chief said that spectators rushed to the aid of the firemen, helping several of them roll on the ground to extinguish their blazing garments. A call was sent for state police aid and the injured men were rushed to the hospital in state and local police cars and in private automobiles.

Fireman Who Made the Supreme Sacrifice

Francis P. Dunnigan, 53, a fireman and former chief of the Westport department.
Injured Firemen

John Gallagher
Irving Samitz
John Saviano
John Powers
George Powers
Dominick Zeoli

Chief Dolan said that an investigation would be launched today and that State Fire Marshall Edward J. Hickey would also institute a probe into the cause of the explosion.
The Kingston, *Daily Freeman*, New York, **1946-05-03**

Greenville, SC, May 20, 1946, Propane Gas Explosion

Armed soldiers today patrolled littered streets where six persons were killed and over 150 injured here last night when exploding propane gas demolished a block-long Ideal Laundry and wrecked or damaged every building in the vicinity (Figure 2.144). The troops from the Greenville

Figure 2.144 James E. Meyers of the American Red Cross took this photo of the wreckage six persons were killed and over 150 injured here last night when exploding propane gas demolished a block-long Ideal Laundry. (Greenville News File Photo.)

Army Air Base were ordered to the scene to prevent looting from stores and houses that were shattered by the terrific blast.

The explosion occurred around 6 p.m. Rescue parties who stumbled through the shambles of wreckage to bring out the injured said it was a miracle that the casualty figures were not much higher. While firemen and police were still investigating the disaster, reports were received of a second explosion in a sandwich shop about 15 miles out of town on the highway to Augusta. Three persons were hurt, one critically. Cause of this accident was unknown.

After a survey, the Red Cross reported 15 houses destroyed, and 45 damaged. Three buildings in the area became hollow shells that will be torn down as a safety measure. Eight others were damaged. Hundreds of window panes were splintered for blocks around.

The laundry itself was leveled as if it had taken a direct hit from a 2000-pound demolition bomb.

Across the street the Third Presbyterian Church was badly damaged. Pipes from the organ were lifted from their racks and smashed into twisted metal. An unidentified man and his wife were at dinner with his aged mother next door to the big Ideal Laundry when the first blast came last night.

The concussion buckled the walls of the house and the roof collapsed, burying the older woman. Miraculously, however, all three escaped serious injury. J. Carl Trammell, superintendent of the laundry, paid with his life by staying inside the plant directing its evacuation after leaking gas was discovered.

The leakage was found by E. R. Haynie, plant manager. After spreading the warning, he rushed to a fire station a block away for help. While he was explaining things to firemen, they were almost knocked down by the explosion.

The dead were identified as Mary Brown, Jerline Simpson and Mamie Arle. Fire Chief Frank Donald explained that the propane gas, used as cleaning fluid, was kept in a big tank in the laundry. The gas escaped from a leak in vaporized form and, being heavier than air, settled into the laundry basement where it was ignited by the furnace.

Fireman Who Made the Supreme Sacrifice

W. L. Harbin, Fireman

Chester Times, Pennsylvania, **1946-11-20**

Azusa, CA, August 21, 1946, Jet Plant Explosion

At least seven persons were killed and more than a score injured today as a terrific explosion of volatile fuels wrecked the Azusa plant of Aero-Jet Engineering Corp.

The blast, originating in a mixing plant, jarred the Los Angeles area, and was felt within a 30-mile radius. The plant manufactures jet assisted takeoff (JATO) units for Navy aircraft, powerful explosives designed to give heavy bombers and fast fighters additional power on takeoffs by rocket-like thrusts.

Because of the danger attached to manufacture, of its products. Aero-Jet built its plant around several structures instead of into one building. Fire, spreading quickly after the explosion, caused considerable damage to the property. Sheriff's officers said the seven dead were civilian employees and that there might be other victims in the wreckage. The fire was quelled within an hour, equipment from neighboring cities being pressed into immediate service. Some of the injured were taken to Monrovia, Temple City and Covina.

Navy officers from the bureau of aeronautics took charge of the plant. Many of the products of the corporation are classified by the armed services as secret. Officers of the 11th naval district at San Diego said no naval personnel was among those killed but that one Navy man was injured. The Navy sources said the fire started in the mixing room where fuels are prepared for jet engines and spread quickly to other combustibles.

Charleston Gazette, West Virginia, **1946-08-22**

Bristol, TN, March 1, 1947, Filling Station Explosion

An official investigation into Bristol's disastrous filling station explosion, which Thursday killed five men, critically injured another and caused minor injuries to approximately a dozen other persons, will be conducted by two Tennessee deputy fire marshal's. Announcement of the investigation into the blast, which resulted in property damage estimated at nearly $100,000, was made Friday by Mayor Fred V. Vance who said he had requested it in the hope that "finding the cause may go a long way toward preventing any similar tragedies in the city."

Mayor Vance said he had been told by James M. McCormick, Tennessee Commissioner of Insurance and banking, that Deputies Virgil Kitts and Tyler Greene, of Nashville, have been assigned to ascertain the cause of the explosion.

Meanwhile, Fire Chief Walter A. Buckles, of Bristol, Tenn., said that the explosion at Robinette's Service Station could have been caused by only one thing. "Gasoline fumes had leaked into the basement and in some manner became ignited," he explained. "How the fumes became ignited, of course, is a matter of supposition, but it is my belief that a spark from an air compressor motor in the basement was the cause. It could not have been caused by a lighted cigarette."

"The explosion did not come from the gasoline tanks. Heavy concrete, which rained on the tanks, broke them and allowed the gasoline to pour out, feeding the flames, but the tanks themselves did not explode."

Fire Chief Hugh Worley, of Bristol, Va., the other half of this twin city, said that three gasoline storage tanks were located in the basement of the filling station, unburied and without proper provision, for ventilation. He said the air compressor motor, which probably had been started when the service station operator filled a tire for a customer, was an old brush-type motor, which frequently throws off a strong spark. Normal ventilation may have been closed off by an accumulation of snow and ice, he added.

Both fire departments began a thorough investigation of all other service stations in the city Friday. City ordinances forbid the type of gasoline storage used at the demolished station, but Robinette's property was built before the ordinances were enacted, a city official said Friday.

Chief Worley said that the life of one of the men, still alive when firemen arrived might have been saved if the fire departments had possessed foamite generators to blanket the flames. Valiant firemen of both departments attempted to rescue the men pinned under the wreckage Thursday before the scorching flames were quelled by water, which had to be used in the absence of foamite, considered far more effective in gasoline fires. "It was awful. It was awful," Jasper Richardson, 41, a customer at the station when the blast occurred, murmured from his hospital bed here Friday. Richardson, a thread mill employee, was the only person who was in the service station or its areaway to survive the blast. He was rescued from the ruins 20 minutes after the explosion by firemen and rushed to King's Mountain Memorial Hospital where his condition is still described as "critical."

Richardson was unable in brief periods of consciousness Friday, to remember the events which immediately preceded and followed the explosion. Whispering questions of "what happened?" to relatives around his bed, he asked if he had been in an automobile accident.

"Something blew up and I can't find my car keys," was the closest he came to an account of the A thorough check of the rubble that once was a masonry building convinced searchers that there were no additional deaths from Thursday's explosion.

Kingsport News, Tennessee, **1947-03-01**

Waltham, MA, March 7, 1948, Plastics Plant Explosion

Two men were killed, two are missing and 29 others injured when a midnight explosion that could be heard for ten miles leveled a plastic factory.

The bodies of the two workmen were removed from the wrecked Interlake Chemical Corp. two-story cement and stucco plant late today as rescuers worked with a power shovel to clear the rubble.

Both bodies were so badly charred they could not be identified immediately.

Search for the two missing men was called off shortly after nightfall. Police and fire officials said they would resume operations at daybreak tomorrow. The explosion and fire left the sprawling plant of the makers of "Makalot" plastic dishes and novelties a seared mass of girders and heavy machinery piled eight feet deep.

Fire Chief B. A. Neal said apparently most of the survivors were blown clear of the wreckage, some of them 50 feet in the air. He estimated damage at "about $150,000."

Raymond Baker, JR., 25, one of the injured, said through flame-seared lips: "I still don't know what happened. There was no warning, like a rumbling or anything like that. Just a terrific noise. The side of the office blew out. I found myself on the floor and crawled outside." "I don't know what caused the explosion. Perhaps it was plastic dust."

Titusville Herald, Pennsylvania, **1948-03-08**

Sioux City, IA, December 15, 1949, Ammonia Plant Explosion

Amid a scene of ruin "just like the place had been bombed," workmen and machines today dug through rubble where at least 16 persons perished in a violent explosion (Figure 2.145) yesterday.

As sorrowing families of the 16 identified dead went about funeral preparations the search at the Swift & Co. packing plant continued for three persons still missing. Nine of some 90 other persons injured remained in critical condition at hospitals.

Leaking gas which had hampered search operations was stopped last night.

Figure 2.145 Amid a scene of ruin "just like the place had been bombed," workmen and machines dug through rubble where at least 16 persons perished in a violent explosion. (Sioux City Public Museum.)

Workmen who labored throughout the night under floodlights used blow torches to melt away twisted girders.

The searchers were able to discard their gas masks after ammonia fumes from the plant's shattered refrigeration system were cut off at their source. The blast rocked the building shortly before the noon hour yesterday. There were about 1,000 persons in the building. The front end of the building was hit hardest. The first and second floors, which housed the office staff and company restaurant, got the brunt of the blast. Part of the second floor ceiling caved in. A reinforced concrete loading dock at the southwest corner of the building fell into the basement. Several bodies were recovered there.

There was an almost warlike atmosphere today as trucks and bulldozers moved ahead on the slow job of removing the debris. A veteran of the European front in World War II said he had seen lots of bombed out buildings "but never anything as bad as this." National Guardsmen still patrolled the area around the plant. It is located in the heart of the Sioux City Stockyards, one of the nation's major terminal markets.

No formal investigation into the specific cause and location of the explosion had been begun. Fire Chief Charles S. Kuhl assigned escaping natural gas as the cause. Mayor Dan J. Conley, who called a meeting of all key officials for today "to go over the whole situation," said an investigation "definitely would be made." The direction of his investigation, he said, remains undetermined.

At present, the mayor continued, the search effort remains in charge of the Iowa National Guard units at the scene It seemed unlikely any mass funeral would be held for the victims. Private arrangements for last rites already had been made by two families.

A Swift & Co. spokesman said the plant would be shut down "for some time."

Assistant Fire Chief Clarence Cappelle said last night he expected to find four more bodies in the debris. This figure was reduced to three today after one of the four missing persons reported he was safe. However, H. C. Watts, chief construction engineer for the company said shortly before noon that searchers now doubt that bodies of the three missing persons are still in the plant.

The company said 12 of the persons killed in the blast were employees in the operating departments of the firm and that a company nurse was the only office employee injured fatally. Three others killed were outsiders at the loading dock.

Twelve company experts, headed by K. H. Clarke, vice president in charge of plant operations, began a study into the specific cause of the explosion. A company spokesman said "it may be some time—if ever— before an exact cause is determined." He said the plant had been using natural gas since 1932. Representatives of the firm's Employees Benefit

Association were calling on families of victims to see what financial aid is needed. The Red Cross also announced it has funds on hand to meet emergency needs.

Fire Chief Kuhl, estimated the damage "at more than $1,000,000." A spokesman for 12 company officials, who came here from Chicago by a chartered plane, said they were conducting an investigation to determine the cause of the explosion. He said the officials "couldn't even guess" the extent of the damage.

Moberly Monitor – Index, Missouri, **1949-12-15**

Crossville, IL, January 14, 1952, Propane Gas Explosion

Two propane gas blasts hurled pieces of a large metal tank into homes over a two-block area last night and killed an elderly widow. Two other persons were injured.

Mrs. Mae Slankard, about 55, died, either directly from the blast or a fire which demolished her house. Seven other frame houses were damaged badly in this small Southern Illinois town.

Police Chief R. G. Randall said the first blast ripped apart a 500-gallon tank of the gas. A second explosion, less than a minute later, occurred in the Slankard house. Randall said he didn't know what touched off the blasts.

A large section of the tank ripped through Randall's kitchen. "We didn't have time to think what it could be," Mrs. Randall said. She was in the dining room with her three children. She ran outside where she found Mrs. Charlotte Cox, telephone operator. Both women ran to the nearby telephone office and spread the alarm. Mrs. Allen Cozard, about 65, was burned critically. Her husband was injured slightly.

The tank belonged to the Rev. C. C. Campbell, Baptist minister. He said he and his family were getting ready for church. "The first explosion knocked my son Charles against me and almost tipped me over." They ran outside, uninjured. The large section of the tank that went through the Randall kitchen flew over two vacant lots, another house, the telephone exchange, and slammed against the second story of another house.

Daily Times-News, Burlington, North Carolina, **1952-01-14**

Houston, TX, June 6, 1953, Fireworks Plant Explosion

More than 40,000 pounds of 4th of July fireworks exploded with horrible fury today, smashing a city block. Police said at least four were killed and 75 injured. Spouting smoke and a ball of flame, the warehouse that was jammed with the holiday explosives disappeared with a thunderous roar that was heard for miles. A mushroom cloud of white smoke—like those created by atomic explosions—billowed thousands of feet into the air.

The warehouse was less than two miles from downtown Houston and in a thickly populated area of homes, buildings and apartment houses. It contained an estimated 40,000–50,000 pounds of display fireworks. A spark from a hammer may have caused the blast.

Five hours after the explosion police withdrew an earlier announcement of 10 known dead. Rescue workers said additional bodies may be found in the wreckage of buildings, homes and apartments.

Every building in the block surrounding the warehouse was either leveled or severely damaged. Residents as far away as five miles felt the blast. City Hall and newspaper telephone switchboards were jammed as people sought to find out what had happened. Many who had seen the mushroom cloud could think only of the pictures they had seen of atomic explosions.

The blast occurred at mid-afternoon at the Alco Fireworks and Specialty Company warehouse.

All that remained of the 40 by 60 foot warehouse were twisted pieces of steel and corrugated iron (Figure 2.146). An eight-foot steel beam was found a block and a half away. The four known dead were found in the wreckage of a small cottage which stood only 20 feet from the warehouse. All four were burned practically beyond recognition.

Three were of the same family. They were Mrs. Jean Walton, 25, and her two children, John, Jr., 2, and Cathy, 4. The fourth was a neighbor, Mrs. Jessie Jean Barziza, 22. The husbands, John A. Walton and Gurnde M. Barziza, 24, were near collapse at a hospital. Walton is a department manager at a clothing store. Barziza an auditor at a downtown hotel.

The injured were rushed to four hospitals. Most were treated for minor burns, cuts and bruises. Only a few were reported to be in serious condition.

Figure 2.146 All that remained of the 40 by 60 foot warehouse were twisted pieces of steel and corrugated iron. (Houston Fire Department.)

Chief of Police L. D. Morrison said the over-all damage will probably be more than $3 million. Kenneth B. Williams, 35, general manager of the fireworks company, was hammering nails into framework around a pressed-paper display cylinder containing explosives. "I either hit the gerb (cylinder) with the hammer or struck a spark off one of the nails," Williams told Fire Investigators Alcus Greet and Harry Foster.

The cylinder suddenly went into flames, he said, and he dropped it when it burned his hand. Williams said he ran out of the building, followed by another employee, the only other persons in the warehouse. "The whole place was blown up in 15 seconds," Williams said.

Every available ambulance sped to the scene, about two miles west of downtown Houston. They took the injured to four different hospitals. A reporter said the explosion sounded like a car of ammunition blowing up and he could "hear stuff popping like bullets."

Seventy-five police cars patrolled the area to prevent looting from damaged business houses and to control traffic.

Witnesses said the plant blew up with a "perfect atomic mushroom cloud," the white smoke billowing several thousand feet. A "ball of fire" hovered under the top of the cloud several minutes. The warehouse was completely destroyed by the explosion and fire. Three walls of a restaurant commissary next door collapsed. Homes and business firms in an immediate four block area were damaged severely. Buildings as far away as two miles received minor damage.

The explosion occurred at 3:05 p.m. (CST) at the plant site, two miles west of downtown Houston.

Residents from across Houston swamped police headquarters and newspapers with telephone calls. It was five minutes before police could spot the actual scene. By the time the first police car arrived, private automobiles were rushing the injured to hospitals. Most of the galvanized iron roof of the Mason Automatic Vending Co.'s building to the south of the fireworks warehouse was blown away.

J. S. Mason, the owner, said several of his employs were injured but received only minor cuts and bruises. Wreckage from the buildings was found in a street two blocks away. An eight-foot steel beam apparently from the warehouse, landed a block and a half away.

Galveston Daily News, Texas, **1953-06-06**

Marianna, FL, August 14, 1953, Explosion and Fire Bottled Gas

Seven persons were injured, two buildings were destroyed and several others were damaged by an explosion and fire which occurred in the heart of Marianna's business district, about 7:30 this morning. Marianna Drug Store and Smith's Fashion Shop were leveled by the blast and resulting flames.

Gulf Life Insurance Building, a two-story structure, and Turner's a ready-to-wear establishment, both located adjacent to the destroyed buildings, were severely damaged by the flames. Other downtown buildings as far away as a block sustained plate glass damage.

Those most severely injured were identified as Charles McDonald, a Pensacola salesman for the Lyons Drug Company, of New Orleans, who was in the drugstore at the time of the explosion; Lawrence Guess, a Marianna insurance agent; and George Johnson, a store janitor.

All three were hospitalized at Jackson County Hospital but their injuries were not believed to be serious.

J. C. Smith, owner of the Fashion Shop, suffered slight injuries. A fifth person, Floyd Harris, manager of F. C. Daffin Company, a building supply firm, located about half a block from the scene of the blast, suffered slight cuts when he was struck by flying glass as he stood in front of his office building. The explosion shattered glass windows in the second story of the Daffin building and tore off Harris' eye-glasses.

Fire departments from Panama City, Sneads, Chipley and Dothan, Ala., and at least seven members of the State Forestry Service were rushed to the scene to aid in the battle against the fire. At about 9:30 a.m.—two hours after the explosion—the fire was reported under control. The cause of the explosion had not been determined, but it is believed the blast was set off by a tank of bottled gas located in the drug store. The Associated Press quoted Harry Bell, Marianna city and fire commissioner, who operates a grocery store not far away, as saying, "It looked like a bomb had been dropped."

Sonny Harris, of radio station WTYS, said, "It just disintegrated. It was here yesterday and it's not here now." The drugstore had not opened for business when the explosion occurred, but several persons were standing nearby. Some of them were cut by flying glass and debris.

Parts of the drug store were blown across the street. Plate glass windows were shattered all along the street, including those of one store— Chipola Drug Company—located eight stores down the street.

The Marianna drugstore and the Fashion Shop were one-story brick buildings about the center of the block. No estimates on total damage could be obtained this morning, but Lucien W. Watson, owner of Marianna Drug Company building, estimated his minimum loss would be $35,000. Mrs. J. C. Smith, wife of the owner of the Fashion Shop, said their complete stock was destroyed. She said the loss would be "heavy" but she could furnish no against the raging flames which by Miss Mildred Conley.

City employees, from all departments, were thrown into the fight against the raging flames which followed the explosion. Units of Florida Highway Patrol and local law enforcement agencies were stationed at strategic points to keep traffic moving. The damaged buildings fronted on

Lafayette street with their rears toward Market street. A similar blast and fire last February wiped out nine buildings.

Panama City, *News-Herald*, Florida **1952-08-14**

Point Pleasant, WV, December 22, 1953, Gasoline Barge Explosion

"One tragic event that stands out in Marietta's history occurred on the morning of December 22, 1953. On this day an empty gasoline barge exploded at the company's boat dock. This accident claimed the lives of six workers who were cleaning the barge at the time it exploded. The workers were Raymond Beller, Chester Elliot, Sherman Randolph, John Wheeler, William Fletcher and Harry "Tad" Bonecutter." (Guide to the Marietta Manufacturing Company Records)

Philadelphia, PA, October 10, 1954, Chemical Plant Explosion

Three ranking fire department officers and 7 other officers and firemen were killed and 24 other firemen and policemen injured Thursday in an odd explosion confined entirely to the rear yard of a north Philadelphia chemical manufacturing plant (Figure 2.147).

The violent blast, without any fire, occurred minutes after a telephoned warning summoned the fire company to investigate the source of escaping fumes, thought to be ammonia.

A 15-foot high steel tank mounted on a wooden platform in one corner of the yard exploded without warning as firemen searched for the source of the fumes

Figure 2.147 Three ranking fire department officers and 7 other officers and firemen were killed in an explosion at a Philadelphia Chemical Plant Explosion. (Philadelphia Firemen's Hall Museum.)

So powerful was the explosion that it slammed the firemen, and several policemen who stood nearby, with terrific force against two 25-foot-high brick walls enclosing the yard.

Rescuers found the dead and injured lying grotesquely on the ground. The steel tank was torn in pieces. Several windows in nearby buildings were shattered.

Blast Comes Without Warning. The firemen, led by News, were called to the two-story Charles W. at 5:56 a.m. (EST). Outwardly there appeared to be no sign of trouble. The blast came 10 minutes later. Deputy Fire Commissioner George E. Hink said the cause of the fumes and the blast are a mystery. "We searched the plant and we don't know where the fumes came from," Hink said. "There was no damage inside the place. We also have no explanation for what caused the tank to explode. There will be a thorough investigation."

Daily Chronicle, Centralia, Washington, **1954-10-28**

Firemen Who Made the Supreme Sacrifice

Deputy Chief 2 Thomas A. Kline, 59.
Battalion Chief 3 John F. Magrann, 61.
Battalion Chief 6 John J. News, 61.
Firefighter Joseph J. Bandes, 54, Engine 2.
Firefighter James F. Tygh, 33, Engine 29.
Firefighter James E. Doyle, 32, Engine 29.
Firefighter Thomas W. Wilson, 36, Engine 29.
Fire Lieutenant Charles C. Holtzman, 30, Ladder 3.
Firefighter Joseph J. Vivian, 32, Ladder 3.
Firefighter Bernard Junod, 32, Engine 2.

Andale, KS, August 27, 1957, Farmers Coop Fire

Fire, sending smoke billowing from oil storage facilities of the Andale Cooperative Assn., threatened the business district at Andale, 21 miles northwest of Wichita, today. Officials said the blaze apparently started about 11:20 a.m. as a man was loading a gasoline truck, and spread quickly at the co-op, which is at the south edge of the town of some 400 population.

Fireman sought to keep the blaze from spreading to a large tank of highly explosive gas (butane). If it let go, they said, damage to a large part of the business district could result. Smoke from the fire was clearly visible 15 miles distant. Officials said chemical equipment was rushed to the scene from McConnell Air Force Base, Wichita, and the Hutchinson Naval Air Station. Three engines and two tankers from stations of the Sedgwick County Fire Department also rushed to Andale.

A spectacular fire that "could have wiped out our whole town if the wind had been from the south" destroyed three large gasoline storage tanks, a warehouse and a gasoline truck at the Andale Farmers Co-op in nearby Andale Monday. The co-op manager, N. B. Schmitz, said the fire – the co-op's first in 20 years of operation – caused losses totaling $25,000.

The danger to the community of 400 residents, 21 miles northwest of Wichita was told by its 37-year-old mayor, O. H. Peltzer. The mayor paid tribute to the fire departments of at least seven neighboring communities who joined Andale's 18 volunteers in controlling the blaze.

At their height the gasoline-fed flames sent a tall column of black smoke billowing in the air, visible for 15 miles or more. A strong wind, instead of the light breeze prevailing, could have spread the fire into the co-op's elevator string with a million bushel grain capacity. Mayor Peltzer said.

Saved also were a warehouse and six other storage tanks of diesel fuel and gasoline some 30 feet from the blazing area. Loss included the three tanks and the truck, 27,000 gallons of gasoline, and a 30-by-50-foot warehouse containing 120 thirty-gallon drums of oil and a thousand pounds of greases.

Salina Journal, Salina, KS, **27 Aug 1957**

Leonardo, NJ, May 23, 1958, Nike Missile Explosion

Investigators searched a Nike base near here today in an effort to learn what caused eight fully armed missiles to blow up in a furious mushroom of fire and death.

The explosion yesterday killed 10 persons and scattered explosive warheads across a wide area of the countryside. The disaster, described by a general as an accident that could not happen but did, was set off by a single missile that exploded.

A split-second chain reaction turned the entire area into a flaming pit of destruction that one eyewitness called horrible beyond belief. Mangled bodies and fragments lay strewn about where a moment before men had stood. The disintegration of the victims made it difficult to establish identities of all. Three others were injured, one seriously. The dead included six soldiers and four civilians.

Two servicemen in a 20-foot-deep pit under a missile's launching pad miraculously survived the holocaust. Staff Sgt. Joseph W. McKenzie, 33, a launcher section chief from Framingham, Mass., stepped from the pit unhurt. His partner, Pfc. Joseph Abott, 24, Grindstone, Pa., was treated for shock and hysteria. The missiles, known as the Ajax type, exploded at about 1:20 p.m. while a team of experts was working on them. They were to be replaced next year by Hercules missiles capable of carrying atomic warheads.

Each of the Ajax missiles carried three conventional warheads of explosives and shrapnel. Most of the explosive devices were accounted for, but some had still not been located today. Duff said ordnance experts had found that all of the eight missiles had left the launching area, flying various distances.

A 12-foot section of one missile landed in a back yard three-quarters of a mile away.

Patrolman Daniel Murdoch, one of the first at the scene, told of "the horror of seeing men, their bodies still afire, and the head of at least one of the men blown away by the force of the explosion." The Army flew in three inspectors from the office of the chief of ordnance in Washington to investigate the explosion. What set it off may never be known.

Residents of this area had protested in vain against erection of the installation 18 months ago. The Army had told the public no such accident was possible and that the missiles would only be fired in case of war. Windows were shattered and doors blown in a mile or more from the explosion scene. One woman was blown out of a chair in the living room of her home.

The Ajax, about 32 feet long and a foot in diameter, weighs about a ton and is designed to bring down enemy aircraft at altitudes of up to 60,000 feet. It has a range of 15 miles.

Chester Times, Pennsylvania, **1958-05-23**

Orwigsburg, PA, June 3, 1959, Propane Truck Explosion

Five marshals sought today the cause of a gas truck explosion that killed 11 persons and injured 10 others. One victim remained unidentified. The blast Tuesday turned the liquid propane-filled steel tank of the truck into a flaming projectile that hurled spectators and firemen hundreds of feet. It was set off when the truck caught fire after a highway collision (Figure 2.148).

The victims in some cases were decapitated and torn by the flaming flight of the tank, ripped jet-like from its chassis by the blast in 7,000 gallons of liquid propane gas. The tank, spitting bits of metal like shrapnel, landed 150 yards away. The explosion occurred 30 minutes after the Sun Gas Propane Co. truck halted behind a school bus picking up a passenger and was rammed in the rear by another truck on rain-slick Route 122, near Orwigsburg.

About 85 school children in buses ahead and behind the two stricken vehicles escaped injury when their alert drivers sped from the scene, one cutting across a field. Two bodies were hurled against the windshield of State Trooper Earl Klinger's automobile parked 300 feet from the explosion site. Klinger said he saved himself by "ducking under the dashboard."

Figure 2.148 The Orwigsburg blast turned the liquid propane-filled steel tank of the truck into a flaming projectile that hurled spectators and firemen hundreds of feet.

Representatives of the Interstate Commerce Commission joined fire marshals in a study of the accident as police continued efforts to identify the 11th victim of the disaster. He was described as a heavy-set man in a red and black checkered shirt. Five of the 10 injured were reported in serious condition. The injured included Joseph Wharton, 48, Peckville, driver of the propane truck, and Walter Williams, 54, Reading, operator of the tractor-trailer which rammed it.

The identified dead were:

Sgt. Michael Wisniewski and trooper John Ripka, marshals from the state police barracks at Reading, conducted the investigation. Representatives of the Interstate Commerce Commission also made a study of the accident. They were John Barry, Philadelphia, and Kenneth Davis and James Jeffrey, of Scranton.

Firemen Who Made the Supreme Sacrifice

Fireman Earl Hillibush, 77.
Fireman Clifford Kriner, 40.
Both of the Friendship Hose Co. at Orwigsburg.

Lebanon Daily News, Pennsylvania, **1959-06-03**

Speedway, IN, December 4, 1959, Air Products Plant Explosion

Two men were killed and another was injured today when an explosion shook a building of the Linde Air Products Co. plant in suburban Speedway. The explosion occurred shortly after 10 a.m. EST in a department where between 15 and 20 men were at work. Plant spokesmen indicated the three men were testing a flux powder dispenser when the blast happened.

The dead were Lloyd Lawrence, Indianapolis and Joseph Buydos, 30, Speedway.

Injured was Wilbur C. Money, Speedway. Lawrence was an employee of the American Casting Co., Buydos was a Linde production planner, Money is a Linde inspector. Speedway patrolman William Berry said Money was able to give him an account of the explosion. He was apparently not hurt badly.

Money told Berry the three were testing flux dispenser tanks which are supposed to hold pressure up to 100 pounds to the square inch. He said the regulator on the compressed air equipment let too much air in and the top flew off. Buydos and Lawrence were killed outright.

The explosion occurred in a steel tank three feet high which stands on legs and has a pressure cooker-type cap on top and gauges indicating the air pressure.

The Linde plant is a division of Union Carbide & Carbon Corp. It makes oxygen tanks, acetylene tanks and welding equipment. The blast was the third at the firm in three and one-half years, and the second this year. No one was hurt in a 1958 explosion but two employees were injured in a blast last May 14. Buydos and Lawrence apparently were bent over the tank when it exploded in their faces. The brass lid popped 40 feet into the air and hit a steel beam.

Linton Daily Citizen, Indiana, **1959-12-04**

Bayonne, NJ, December 29, 1960, Propane Explosion

A storage tank containing potent liquid propane exploded Wednesday night. Four men were injured, one of them critically, and hundreds fled their homes. Scene of the explosion was the Sun Gas Products Corp., located in Bayonne's "Constable Hook" section, near New York Bay. Flames climber 1,000 feet into the sky and windows for blocks around were shattered by the impact. Police evacuated everybody from within a five-block radius of the explosion scene and cordoned it off.

Firemen sprayed the flames and wet down a huge tank containing oil, just 200 yards away. At one point, flames were within 50 yards of the oil tank. But shortly before 2 a.m. today, firemen announced the fire was under control. Critically injured was truck driver Stanley Buchalski, 40, of Chester, N.Y. He was taken to Bayonne Hospital suffering from third degree burns over 40 percent of his body.

Also injured were Billy Rowan, 27, of Bayonne, a volunteer worker; Earl Lewis, 47, of Port Jervis, N.Y., a truck driver; and Harold McCarter, 39, also of Bayonne. These men were not seriously injured. Police retracted an earlier report indicating one man had died, when a missing man was located. Police gave this account of events leading up to the explosion:

Three men, including Buchalski and Lewis, were engaged in loading gas into a small tank. Somehow, a spark ignited the gas. A small drum of gas exploded, then another and another. Buchalski and those with him tried to move their truck away from the exploding drums. It was then that Buchalski was burned. Nobody is sure how many drums exploded; one report indicated there were 18. But the smaller explosions were enough to touch off the big storage tank, which went off with earsplitting force.

Port Angeles Evening News, Washington, **1960-12-29**

Mitchell, IN, June 22, 1961, Lehigh Portland Cement Co. Explosion

Four workers were killed and a dozen others were injured Wednesday as an explosion attributed to an accumulation of natural gas wrecked a nearly-completed building at a cement plant. Rose Ann Blackwell, 54, Mitchell, a cleaning woman working her first day at the Lehigh Portland Cement Co. plant, and Harry May, 38, Bedford, were killed in the blast. Seward Mundy, 55, Bedford, and Larry Runyon, 27, Bloomington, died in Dunn Memorial Hospital.

Reported in critical condition were Griffith Ogden, 34, Indianapolis, Emery Cooper, Bedford, and John Benny Roberts, 27, Evansville. Richard Perkins, 20, Indianapolis, Lloyd Ramsey, 56, Bedford, and Claude Cade, 40, Bedford, were listed in serious condition. Six others required medical treatment but were not hospitalized. Several suffered slight injuries.

Trooper J. K. Miller, of the Seymour state police post, was sent to the scene of the explosion at Mitchell Wednesday afternoon to assist with handling sightseers during clean-up operations. Much help was needed in the area after the blast. Mrs. Blackwell had just started on the cleanup job in the final stages of construction of the building, a scale house. The men were construction workers who had been building the 60 by 90-foot building, where trucks were to be weighed in and out of the plant.

"There's no question but what it was a natural gas explosion," said David Volk of Bedford, project engineer for the Fruin-Colnon(sic) Construction Co. Volk said natural gas was being piped into the building for heating purposes. "We're very upset about it and we're doing all we can to ascertain what the exact cause was," Volk said.

The explosion occurred at lunch time. Some of the 24 workers on the construction crew were away and some were just returning. The cement plant employees about 250 workers, but they were in other buildings separated from the scale house. James Hensinger, Lehigh general manager here, said the blast occurred after a workman struck a match preparatory to lighting two gas water heaters. The workman believed to have been one of those who died in the explosion. Hensinger said the blast blew the roof and other parts of the building 100 feet in the air and collapsed all four walls.

Daily Tribune, Seymour, IN newspaper, Seymour, IN, Thursday, **22 June 1961**

Berlin, NY, July 26, 1962, Propane Gas Explosion

At 5:30 p.m. in the picturesque hamlet of Berlin in eastern New York, dinner was on the stove, the men were returning from work, and Gene Merrills was standing beside a barn.

"God, it was like a clap of thunder. A bomb. Everything lit up red," Merrills said.

A load of propane gas on a tractor-trailer exploded Wednesday, spurting the flaming liquid as far as a half-mile within this community of 490. The truck driver was injured fatally and 16 men, women and children were hospitalized. Ten of them remained in critical condition today.

A dozen homes and the 179-year-old Baptist church were in ruins.

Mcucas, 39, of Pomeroy, Pa., took his big truck down the twisting two-mile hill leading into this community folded into a valley between the Grafton Mountains and the Berkshires near the intersection of the New York-Massachusetts-Vermont state lines. At some point, the brakes apparently failed, State Police said, and McLucas leaned on his horn in warning as the vehicle picked up speed. Some witnesses said the brakes appeared to be on fire.

About 50 yards from the village square with its Civil War statue, the truck reached a turn near the bottom of the hill. It jackknifed, the trailer broke off (Figure 2.149). "There was a pause, then a pop, than a bam.

Figure 2.149 About 50 yards from the Berlin village square with its Civil War statue, the truck reached a turn near the bottom of the hill. It jackknifed, the trailer broke off. (Times Union Photo.)

The flames must have gone 150 to 200 feet in the air bright orange." That was the account of Holden Gutermuth, 24, who witnessed it from a field. McLucas was blown 350 feet. He died about two hours later.

Postmaster Robert Moses was closing up for the day. He looked at the house he had built himself. "It was terrible. It was a mass of fire." His wife was inside. Held back by the searing heat, Moses watched as his wife crawled into the breezeway, then through the garage, over the car and through a window. She was critically burned.

Three homes were enveloped immediately in flames and hunks of fire soared off to strike in a crazy-quilt pattern among the few dozen white clapboard homes in Berling and the church. A barn a half-mile away housing two school buses was ignited. A man painting his house a quarter-mile away was blown from his ladder. The immediate blast occurred in front of the home of Kenneth McCumber, 53, and his wife, Florence, 55. The house disintegrated.

"You could see the clapboards come right off the place, said Clifton Shuhart, who found the McCumbers crawling in a field about 400 feet away. They, too, were burned critically."

Next on the hill was the Brazie home. Inside were John Brazie, 23, his wife, Mary, 18, and their son, John Jr., 18 months. "I saw a woman and baby run out of a house. They were all afire. It was horrible," said Miss Leona Gayle Jones, 23, who was driving home from her state job in nearby Albany. The Brazies were in critical condition.

Fire Chief Frank Jones responded to the alarm and found his own home burning. What he could not find was his wife. "He's in terrible shape," the Rensselaer County fire coordinator told a newsman at the scene. "He couldn't find his wife. He's just in a daze." But Mrs. Jones was found uninjured at a neighbor's home. Sixteen fire departments from communities in the county and from Massachusetts fought the flames. At least 11 ambulances, 21 state troopers, and 9 sheriff's deputies responded.

Eight of the injured were in Putnam Memorial Hospital, Bennington, Vt., six were in Samaritan Hospital, Troy, N. Y., and two others were discharged after treatment in Troy.

Nashua Telegraph, New Hampshire, **1962-07-26**

Dunbar, PA, February 22, 1966, Fireworks Plant Explosion

Five women were killed and eight other persons injured when an explosion ripped the Keystone Fireworks Co. plant in Dunbar this morning. Four were killed instantly and a fifth woman, Mrs. Freda Bezilla of Dunbar, died in nearby Connellsville Hospital of injuries received in the blast. The five victims were working in a building which produced "cherry bombs"—a noisy firecracker used in fireworks displays. The building was completely leveled.

A sixth employee in the building—June Martin Orawiec of Dunbar—was reported in "critical" condition at Connellsville Hospital. The other injured employees were at work in other buildings in the complex located on a hill above Dunbar. Authorities said one of the first persons on the scene was Fireman Lou Ross of Dunbar, who found his wife among the dead.

Admitted to Connellsville Hospital for treatment were Mrs. Marion Tressler and Mrs. Mildred Miles, both of Dunbar.

Treated and released were Mrs. Myrtle Balsley of Uniontown; John Ruse, Dunbar; Mrs. Lilly Peck, Lemont Furnace R.D. 1; MRS. Lucy Roebuck, Dunbar R.D. 2; and Mrs. Hazel Bartlett, Dunbar. The owner of the plant, 74-year-old Bedy Lizza of Dunbar, was also admitted to the hospital when he went into shock and collapsed. The blast, which occurred about 8:45 a.m., demolished a frame building and damaged others. The force of the explosion was felt as far as three miles away.

The dead had been mixing chemicals when the explosion occurred, authorities said.

The company was in the midst of its busy season, preparing a fireworks inventory for Memorial Day and the Fourth of July. Police said previous explosions had hit the plant, the most recent occurring last summer. That blast was on a weekend, however, and no one was injured. At least seven fire companies went to the scene today. Robert Foltz, a newsman from nearby Uniontown, was one of the first at the site. "The bodies were badly mutilated," he said. "The building itself was leveled and those around it damaged." Dunbar is a community of about 1,500 persons.

Valley Independent, Monesen, Pennsylvania, **1966-02-23**

Dunreith, IN, January 8, 1968, Train Wreck and Explosions

Thunderous explosions from ammonia-filled tanker cars after two freight trains crashed forced a mass evacuation Monday night and set fire to two businesses and several homes (Figure 2.150). All 236 residents of this east-central Indiana community were evacuated safely. Three firemen and a policeman suffered minor injuries. The Butterfield Canning Co. plant and a service station were destroyed by flames, which also spread to 10 houses. Some were empty migrant worker dwellings, officials said.

A westbound Pennsylvania Railroad train derailed and sideswiped an eastbound Pennsy freight. W. R. Sheets of Zionsville, brakeman on the westbound train, said he believed a rail broke under the 75th car of the 98-car train. Flames broke out almost immediately he said.

Officials quickly removed townspeople from the danger area of the chemical laden tankers.

Mrs. Mabel McGuire, who lives a half block from the accident scene, said one explosion "looked like the whole world was on fire."

Figure 2.150 Thunderous explosions from ammonia-filled tanker cars after two freight trains crashed forced a mass evacuation Monday night and set fire to two businesses and several homes. (Henry County Historical Society.)

"Our back door was blown off the hinges," she said. "There were a lot of small explosions and one big one that lit up the whole sky." The blast was felt in areas 25 miles away. A plot said the explosions shook his small plane as he flew over Dunreith, which is about 35 miles east of Indianapolis. Traffic was rerouted from busy U.S. 40 which parallels the Pennsylvania main line through Dunreith.
Emporia Gazette, Kansas, **1968-01-02**

Blakely, GA, January 27, 1970, Butane Gas Explosion

Two persons were killed and six injured today in a butane gas explosion that knocked out virtually every window in the downtown area, touched off a raging fire, and shook houses 14 miles away. "It sounded like a jet plane had crashed," said Police Chief Carl Gilbert. The explosion sent up a huge, white mushroom-shaped cloud and residents of Arlington, Ga., 14 miles away, said their houses shuddered. Small bits of white paper fluttered to the ground in yards a mile from the scene.

Mayor Alex Howell asked the State Patrol to man an around-the-clock watch over the downtown area to prevent looting. The blast touched off a fire at the Davenport Motor Co. two blocks from the town square. Firefighting units from four nearby towns brought the blaze under control several hours later. Dead were Thomas Edward Deal, 36, manager of the Empire Gas Co., and James Herman Clark, 58, assistant manager of the motor company, where the blast occurred. Two seriously injured victims were admitted to hospitals. Four others were released after treatment.

The blast knocked out major telephone lines, isolating the community for several hours until emergency lines could be set up. The mayor said the blast shattered nearly every window in the downtown area of this town of 5,000. Authorities said the explosion occurred while Deal and Clark were investigating a gas leak at the motor firm where two butane gas trucks were being repaired. Deal and Clark were killed instantly in the blast which scattered debris all over the downtown section and shattered windows in about 75–80 businesses. The motor firm's building was flattened, along with a nearby grocery store and the gas company building.

Press-Courier, Oxnard, California, **1970-01-27**

Brooklyn, NY, May 31, 1970, Liquid Oxygen Truck Explosion

A tank truck loaded with liquid oxygen exploded outside a Brooklyn hospital Saturday, killing two men and injuring about 30 people. The blast shattered windows in Victory Hospital and about 20 surrounding apartment houses, shaking some people out of bed on a peaceful holiday morning. All of the injured were cut by flying window glass. Most were patients or staff members at the 117-bed hospital, police said, and several were evacuated to undamaged rooms.

Wolfgang Reich, who had been watching from his apartment bedroom, said it sounded "like a bomb."

Phillip Josephs, 22, who was washing his car about 200 feet away, said he heard a hissing noise and a loud bang and saw flames shoot into the air. He ran down the tree-lined street to turn in the fire alarm at the corner box. Firemen quickly extinguished the blaze. They said they did not know its cause, and believed it to be the first such explosion in the city's history. The dead were identified as Edward Graves, 22, the truck driver, whose clothes were blown off, and Steve Pascale, 57, a Brooklyn plumber, who was in a shack on a construction project near the hospital.

Police said they found a receipt in the truck's cab for 380 gallons of oxygen delivered to Victory Hospital. Officials said such deliveries are routine and that most hospitals and other buildings which use liquid oxygen have outside storage tanks. The truck was owned by Liquid Carbonic of Paterson, N.J., a subsidiary of General Dynamics Corp., police said. It hold 2,800 gallons, and GRAVES had planned to make deliveries to other hospitals. Mayor John V. Lindsay ordered a full investigation. In an unconnected incident several hours earlier, an explosion ripped a two-foot hole in a trailer at the World Trade Center, under construction in lower Manhattan. Police said it was a bomb. No injuries were reported.

Syracuse Herald Journal, New York, **1970-05-31**

Hollywood, FL, September 9, 1971, Dynamite Truck Explosion

A truckload of dynamite blew up at a rock pit Thursday, apparently killing three men. Two bodies were found and the man handling the dynamite was missing without a trace. A spokesman for Broward

County Sheriff Ed Stack said the cause of the blast had not been determined, but that it appeared a dynamite charge being either placed or removed by a blaster's assistant identified as McKinley Jones went off accidentally, setting off an estimated ton of dynamite on a nearby pickup truck.

No sign of Jones was found around the hole blown in the ground by the blast, which destroyed the truck which carried the explosives and overturned another nearby truck and a drilling rig being used at the quarry. The bodies of Curtis Lee Smith and William Flateau were found near the blast scene. Flateau was the driver of a tool company truck delivering drill bits to the quarry and Smith was a dynamite blaster standing near the drilling rig when the explosion occurred, police said. A fourth man, David Frair, was in another truck near the scene but was only slightly injured when the blast blew out the windshield.

Panama City, *News-Herald*, Florida, **1971-09-10**

East St. Louis, IL, January 23, 1972, Tank Car Explosion

A railroad tank car containing 30,000 gallons of liquid petroleum gas exploded Saturday shattering windows 8 miles away and sending more than 150 persons to hospitals. No fatalities were reported and only 18 of those injured, most of whom were cut by flying glass, were admitted to hospitals. City Fire Commissioner Elmo Bush said a quick survey showed the dawn blast in the Alton & Southern Railway Co. yard near downtown, damaged at least 500 residential and business properties here, in most cases shattering windows.

A spokesman for the rail yard said freight cars and four tank cars containing the liquid gas were being pushed over a hump to provide momentum as they were rolled onto a side track during a sorting process. Retarders on one of the tank cars failed to slow it down he said, and the tank car crashed into a freight car and ruptured. The damaged tank car burst into flames following the explosion, felt as far away as Edwardsville to the north and Troy to the northeast, both 20 miles from the scene.

Cedar Rapids, *Gazette*, Iowa, **1972-01-23**

Sioux City, IA, April 30, 1974, Grain Elevator Explosion

Four men were crushed under tons of concrete and grain here Tuesday when sparks from machinery apparently touched off a dust explosion in the silo at the Terra grain elevator (Figure 2.151).

Authorities identified the four killed as George Gruneich and Martin McCormick, both of Sioux City who were employees at the Bartlett Grain Elevator, and Edward Johnston, 58, of Sioux City and Norbert Bertrand, 50, of Jefferson, S.D. Officials said Johnston and Bertrand were employees of the Sioux City grain exchange.

Figure 2.151 Four men were crushed under tons of concrete and grain here Tuesday when sparks from machinery apparently touched off a dust explosion in the silo at the Terra grain elevator. (Sioux City Public Museum.)

Injured in the blast and hospitalized in fair condition with burns and possible fractures was Dean Fray, 39, of Bronson. Fry, a grain worker, was the first man removed from the explosion site. Authorities said their initial investigation revealed that the explosion which occurred near the top of the silo, was apparently ignited by a spark from machinery. Officers said 11 persons were working at the mill at the time, and that FRY and the four killed were working in the vicinity of the silo where railroad cars were being loaded with corn.

Several rail cars, authorities said, were extensively damaged by the debris. They said the elevator was still standing, but that debris covered a 150 yard radius around the silo.

Flying cement also caused extensive damage to the American Popcorn Co. building located about 100 yards to the west of the silo. A wall was pushed in at the popcorn plant by the impact from the explosion, but no injuries were reported.

The elevator is located in the northwest part of the city along U.S. 75. Authorities said several police and fire units were dispatched to the scene, but that rescue efforts were hampered temporarily while officials feared a second explosion. A witness said the top of the elevator's silo "blew off" and spilled concrete and grain on the five men working below. Three of the bodies were found in the debris and the fourth was discovered in a tunnel under the elevator.

Ames Daily Tribune, Iowa, **1974-05-01**

Decatur, IL, July 19, 1974, Rail Yard Collision Fire and Explosion

GATX 41623 and four other tank cars loaded with isobutane gas were uncoupled at the west end of Decatur Yard by a switching crew and allowed to free roll eastward on yard track 11. The car impacted an empty boxcar, and its coupler overrode the tank car coupler and punctured the tank. Isobutane escaped and vaporized for 8–10 minutes before it exploded (Figure 2.152) The yard, surrounding residences, and commercial facilities were damaged extensively by fire and shock waves. Seven employees died from burns, and 33 employees were injured. Three hundred sixteen persons outside the rail yard were also injured as a result of the explosion (Figure 2.153). Property damage was estimated at $18 million.

Figure 2.152 The car impacted an empty boxcar, and its coupler overrode the tank car coupler and punctured the tank. Isobutane escaped and vaporized for 8–10 minutes before it exploded. (Decatur Fire Department.)

Figure 2.153 Three hundred sixteen persons outside the rail yard were also injured as a result of the explosion. (Decatur Fire Department.)

The switch crew apparently opened the appropriate switches into track 11 and cut loose the five tank cars from the locomotive quite some distance from the center, level portion of track 11. The cars then continued rolling, under their own momentum, toward the east. At this time the switch crew left the track 11 area, and began another switching job to the south. These men expected, as in the past, that the released cars would roll east and couple up with other cars already resting at intervals along the track. Track 11, as others in Decatur, is "saucer" type, that is has a slight decline from both ends toward the center of the track, where the roadbed then levels out for a distance.

The heavy tank cars, possibly released from the locomotive at a speed somewhat above what it should have been, gained considerable speed as they rolled downgrade toward the center of track 11. About 2,600 feet from the release point with the locomotive, the lead, east end tank car struck an empty boxcar. The boxcar jumped up, off the trucks (wheels) at the west end, and the coupler of said car punctured the head or end of the tank car.

Now there were about 5,000 gallons of isobutane flowing from the ruptured tank car each minute, 24% of it vaporizing immediately into a highly flammable cloud, the remainder pooling about the area of the leaking car. The capacity of the tank car was 33,000 gallons, and it held 31,299 liquid gallons net contents when full, before being punctured. About 18,000–20,000 gallons flowed out as a liquid, after which some began to vaporize, adding to the cloud already hanging over the area. As isobutane is over twice as heavy as air, the vapor stayed near the ground.

At three minutes after 5:00 a.m., it went off when some unknown source of ignition torched the mammoth cloud of death blanketing the railroad yards. The exact source or sources of the ignition will never be known, as so many possibilities exist in a railroad yard. Idling automobile engines, diesel locomotives, refrigerator cars, pipes and cigarettes, coal oil lamps on switches, charcoal fires on cabooses, sparks from railroad brakes or wheels on rails, trash fires, coupling-caused sparks, fuses, the possibilities are many.

Witnesses in the area said they heard more than one explosion, possibly three or four. As City Fire Marshal Bob Corey said "there were probably several different explosions throughout the area because of the explosive items of L.P. gas. When an explosion occurred, any pockets of gas that were too rich would simply wait a few seconds until the air mixed and then explode. Several of the rail cars would be separate collecting pockets for the gas to accumulate in."

The fire balls flashed brightly in the infant dawn's breaking light. Many persons, one a fire fighter at number one fire house in downtown Decatur, another the author's father, who was many miles southeast, were awakened by the brilliant light. Others were most unfortunate, being in the midst of the flammable vapors and were instantly burned, several fatally.

Within split seconds a terrific shock wave or concussion reverberated over many square miles, and was reportedly heard by state policemen in Pesotum, 45 miles east. Tens of thousands of windows panes were shattered, and slicing, slashing shards were blasted into and throughout hundreds of houses and other structures within a huge area. Several persons had seen the flash, jumped up to a door or window to investigate, and were cut as the glass tore into them. Window unit air conditioners, window and door frames, siding, insulation, rafters, and pieces of roof also were ripped loose and slammed or fell with great force into many damaged homes.

Smoke was extremely heavy, and scores of fires were raging over a very extensive area about 7/8 of a mile long and 1/2 mile wide. At this time, both off-duty shifts of firemen, the 2nd and 3rd platoons, were called to duty.

Probably the greatest single asset the Fire Department had was the level-headed, calm, calculating individuals who made up the top command structure of the Department. This lead to cautious, reasonable, effective use of all apparatus and manpower. These men were not only in charge of 115 City fire fighters, but also scores of County volunteer fire fighters and their equipment. These County units had been requested only minutes after the first alarm was given, and hurriedly responded to Decatur's aid. The use of several of these Departments portable drafting tanks, which hold 1,000 gallons of water, were invaluable in this fire fighter, and several City engines would have been useless or nearly so without them. These County fire protection districts vehicles also protected the City of Decatur while the City's majority of apparatus was tied up at the blaze. City firemen, one at each fire house, rode with the volunteers to help guide them in event a fire struck at an address unfamiliar to the out of town firemen. This was the City of Decatur's hour of need, and our volunteer friends proved their worth under fire.

The situation at the yards, although handled extremely well, presented awesome and confusing incidents. Directions, locations within the rail yard, and inability of each crew to know where, who, or even if anyone was working in another particular area were limited. officers, especially higher ranking ones, worked very hard in coordinating firefighting crews all over the yards, shifting men from one area of high priority to one of higher priority.

The threat of additional explosions was very real, and this made all aware of their dispensability and the frailty of human life in such a complex set of circumstances. Talk of a three mile evacuation area, which wouldn't necessarily include the fire fighters, carloads of dynamite and bombs, and the picture of desolate destruction before each man made all scared, but the threat was faced and rapidly progress began to show.

When several of the really dangerous spots were secured, about 2 1/2 hours after the initial alarm, the fire fighting became mostly routine. The extinguishment or removal of cars containing toxic, explosive, or flammable materials, especially the four propane tank cars sitting next to the ruptured, burning butane car that began the catastrophe, stabilized the situation and the "mop up" type of fire fighting could proceed.

Scores, more likely hundreds, of fires raged over the approximately three-quarter mile square district. Everywhere one looked he saw boxcars, hopper cars, and other wreckage afire, some blazing brightly, others just smoking, about to burst into open flames. Many cars, seemingly missed by the radiated heat, were pulled out of the area to the west by locomotives and stored far from the main fire theater. Dozens of these later burst into flames, and presented an eerie sight, especially at night, when one could see many cars either glowing or blazing defiantly far removed from the place fire fighters were working.

There is nothing local government can do to prevent similar catastrophes, only to try to cope with them after occurrence, when it may be too late. Likewise, state bodies are limited in their degree of regulation of interstate commerce and transportation. With man's ability to create and invent the thousands of lethal products now being hauled on the nation's highways, railroads, and canals, it is rather remarkable to think he is no master of these when a wreck, fire, or other similar incident leaves him with little technology and defensive weapons for his and his community's safety and indeed, future existence.

Excerpts from Reports Compiled by Decatur Fire Personnel:

Robert R. Corey, Acting Assistant Chief
William B. Turner, Fire Inspector 8/7/74

The National Transportation Safety Board determined that the probable cause of the accident was the over speed impact between the heavy cut of tank cars and the uncoupled light boxcar, which resulted from the release of the tank cars at a higher than acceptable switching speed. The lack of written guidelines to assist the switchman in determining the proper switching speed contributed to the accident. The crewmembers' lack of understanding of the risks involved in switching hazardous materials also was a contributing factor. (National Transportation Safety Board Investigation)

Iowa City, IA, January 24, 1975, Propane Explosion

Two workers were killed and three others were injured in a propane gas explosion and fire Thursday that destroyed a refrigeration unit at the terminal of the Mid-America Pipeline Co. four miles east of here. The victims were identified as Terry Guthrie, 28, of Iowa City, and Tony Bryant, 29, of West Liberty, both employed at the terminal as maintenance men, the Johnson County Sheriff's office said.

Three Mid-America employees remained hospitalized Thursday night - one in critical condition - in the burn unit of University Hospitals here. Most seriously injured with second and third degree burns was Leonard Elliott of rural West Branch, officials said.

Also injured in the explosion was Richard J. Saboruin of North Liberty, listed in serious condition, and Ronald Digney of West Liberty, listed in fair condition.

A roaring ball of flames from the Mid-America terminal - a division of Mapco, Inc., based in Tulsa, Okla. - was reportedly seen as far as 10 miles away after the explosion shook the area about 1:30 p.m. Although the fire raged dangerously close to storage tanks and a loading dock at the terminal, no one in the surrounding area was evacuated.

Firemen said they were aided by southerly winds that kept the flames and smoke blowing away from equipment in the terminal. The explosion has been blamed tentatively on a malfunctioning refrigeration unit installed a year ago at the Iowa City complex.

Gilbert V. Rohleder, a Mapco vice-president, said the "chiller" unit used to keep vapor pressure stabilized failed "for no apparent reason." He said pressure gauges on the unit were normal at the time of the explosion.

The two maintenance men killed in the fire were injecting the chemical methanol into the unit just before the blast, Rohleder said. Firemen were unable to remove the charred bodies of the victims until after the flames subsided.

Des Moines Register, Iowa, **1975-01-24**

Collinsville, IL, August 7, 1978, Propane Tank Car Explosion

The 200 residents who evacuated their homes when a tank car full of propane exploded have been told to stay away from their houses until at least Tuesday. Two persons were injured early Sunday when the tank car derailed, exploded and threatened to explode four other tankers.

Police evacuated everyone within a half mile of the explosion site on the outskirts of town while firefighters attempted to keep the fire from spreading to the other propane cars, three of which also derailed.

Evacuated residents were asked late Sunday to remain away from their homes at least until Tuesday, when the fire was expected to burn itself out. After the area cools down, the propane would be pumped from the remaining tanks. A man injured in the blast, Wilburn Alford, 37, of Troy was reported in serious condition in a St. Louis hospital. Police said Alford, who had stopped his truck at a railroad crossing, was burned when the train derailed near the crossing shortly after midnight and the tank car exploded.

Another motorist, Gina Pendleton, 17, of Collinsville, was parked on the other side of the crossing. She was treated for burns and released. Police said both were alone in their vehicles.

The blast and flash of red light came were seen 40 miles from the scene in west St. Louis County. One caller said the red fireball was "like a little atomic bomb," according to the Illinois State Police.

The explosion burned trees in a wide area around the site, charred railroad ties and a crossing warning sign and melted asphalt at the crossing. After the explosion, police blocked nearby Interstate 55-70 for a brief time as a precaution. Firemen continued wetting down the burning car and neighboring tanks throughout the day Sunday.

Collinsville firemen were assisted by departments from Maryville, Long Lake and Troy. Civil Defense units and auxiliary police helped Collinsville police and Madison County deputies seal off the area and set up a command center in Collinsville. The American Red Cross set up a center for the evacuees at the Collinsville VFW Hall.

David E. Visney, a vice president for the Illinois Terminal Railroad, which operated the 63-unit train, said the cause of the derailment was not immediately apparent. He said more would be learned when inspectors could look at the derailed cars after the wreckage cooled.

The Daily Leader, Pontiac, Illinois, **1978-08-07**

Latham, KS, October 25, 1979, Propane Gas Explosion

A propane gas tank exploded like a "bomb" in Latham's only cafe minutes after closing Wednesday night, killing two employees and nearly leveling an entire block of the small town.

The blast, heard 15 miles away, killed the two women managers of the Latham Cafe, who were alone inside cleaning up, and injured five other people in the southeastern Kansas town of 200.

"It sounded like a loud bomb," said Bruce Reed, 14, who was riding his motorcycle about a block away when the explosion occurred shortly before 7 p.m. "It looked like you would think an A-bomb would look."

Five buildings, including some housing city offices and the post office, were destroyed or heavily damaged. The two women, Carla Hodges, about 40, and Kate Bing, 31, were apparently alone when the propane tank exploded. Authorities said the popular restaurant had just closed for business.

The force of the blast threw Mrs. Bing from the building. She died two hours later at a Wichita hospital. Mrs. Hodges was found in the rubble about 1 1/2 hours after the explosion.

None of the five other people injured were hurt seriously. A woman was hospitalized but was listed in good condition today. The rest were treated and released.

Authorities said a meeting of about 40 members of the Eastern Star women's ground had been scheduled to meet in a community building adjacent to the cafe later in the night.

Windows were reported broken out throughout the town of 200 and at least one witness said the explosion was heard 15 miles away.

Authorities at first were not certain how many people were inside the cafe and searched through the rubble for several hours before determining that everyone had been accounted for.

Investigators from county and state offices were sent late Wednesday to the small town to investigate the cause of the blast.

Logansport Pharos Tribune, Indiana, **1979-10-25**

Raymond, NE, November 18, 1982, Propane Explosion

A propane leak was to blame for an explosion that killed four people at the Raymond Co-Op Elevator, State Fire Marshal Wally Barnett said Wednesday. Barnett said a 1 1/4-inch plastic pipe buried near the elevator and feeding a grain dryer next to it developed a leak. "The ground was saturated from the propane, from one end to the other," he said. He said propane seeped into the basement of the elevator and apparently was ignited by a fuel oil heating stove beneath the office.

The explosion Tuesday blew the roof off the wooden feed-mixing elevator, smashing the business office next door and engulfing both structures in fire. Four people were killed and three were critically injured.

The bodies of Jo King, 42, the company bookkeeper, and Dianne Kissinger, 24, who had gone to the elevator with her husband, were pulled from the rubble Tuesday. The bodies of Don Elenga, an elevator employee,

Figure 2.154 Raymond, NE.

and Earl Nelson, a customer who was buying gloves, were found early Wednesday in the water-soaked debris.

Mrs. Kissinger's husband, Frank, 30; MRS. King's husband, Frank, 47, also an elevator employee; and Dick Krone, 35, the elevator manager, fled the elevator following the blast, their clothes in flames. All were hospital-ized in critical condition. The explosion stunned residents in the farming community of 181 people a dozen miles northwest of Lincoln. Postmaster Janice Kirchoff said she was working across the street when she saw the three men run out of the elevator with their clothes on fire.

Hutchinson News, Kansas, **1982-11-18**

Rowesville, SC, May 25, 1983, Fireworks Truck Explosion

The devastating explosions that killed two people, injured five and damaged homes for miles around Tuesday many have been caused by fireworks loaded in a truck trailer, officials say. Federal, state and local investigators were to return to the scene about 1 1/2 miles north of this Orangeburg County town today to continue trying to figure out the origin of the 3:50 p.m. blasts, which Orangeburg County Sheriff Vance Boone said damaged homes up to 2 1/2 miles away.

"It looked like an airplane came along and dropped a bomb," said county Disaster Preparedness Director William Crapps. The explosions left at least three families homeless, said Boone. The trailer about 250 yards off U.S. 21 may have been used as a "fireworks factory," an investi-gator at the scene who asked not to be identified told The State newspaper.

Crapps said investigators thought the parked trailer was loaded with fireworks and perhaps some dynamite. He said he had been told the blasts may have been touched off by Herbert Branham when he lit a match too near the explosives. Hundreds of cylindrical cardboard casings of the type used to make M-80 fireworks littered the lot.

Orangeburg County Coroner Paul Simmons said there was not enough left of the two bodies to make positive identifications. He said the chest of one of the victims was found more than 250 feet away in a wheat field.

However, Boone said it was believed one of the dead was Branham, who owned a mobile home destroyed by the explosions. "His wife is one of the injured and she has assured us that he was in the buildings," Boone said. Investigators still were retrieving pieces of the two corpses shortly before sundown. Simmons said the remains would be sent to the Medical University of South Carolina at Charleston today for positive identi-fication. Three children were treated and released for cuts and bruises Tuesday, while their 33-year-old mother, Charlotte Branham, remained in stable condition at the hospital this morning.

Aiken Standard, South Carolina, **1983-05-25**

Figure 2.155 Santee, NE.

Santee, NE, August 28, 1983, Propane Gas Explosion

Eight children playing in a cloud of leaking propane gas were engulfed in an orange fireball when the vapors ignited, triggering a blaze that burned for eight hours and forced the evacuation of this Indian reservation community. Ten people were hurt, including a propane truck driver and a woman who lived in a mobile home about 25 feet from the storage tank.

"I heard the explosion and I could see the flames over the rooftops," said Robert Brown, 19, who lives about 50 yards from the scene of Friday afternoon's blast. "And all the children were running around on fire."

The eight children, who ranged in age from 4 to 15, remained hospitalized Saturday with burns. All 400 residents of this town on the Santee Sioux Indian reservation were forced to flee after the 4:30 p.m. explosion, and they returned about seven hours later. The propane leaked as a tanker truck driver, Sam Ruiter, was filling a 10,000-gallon underground storage tank for Santee's public school system, sand Tribal Chairman Rick Kitto, who witnessed the explosion.

He said a crowd of children had gathered as a cloud of vapor formed, about 20 feet high and 50 feet around.

"It was a very hot day, and the gas may have felt cool to their skin," Kitto said, adding that Ruiter had been trying to shoo the children away. Suddenly the gas ignited, engulfing Ruiter and the children in an orange fireball 50 feet in diameter, Kitto said. The truck was quickly backed away by members of the community's fire department, but the 6,000 gallons of propane remaining in the underground tank kept burning, shooting flames 40 to 60 feet in the air.

The children ran in the direction of the fire department, where workers grabbed hoses and sprayed them with water to douse their burning clothes. The children were taken to St. Luke's Burn Center in Sioux City, Iowa. Two were in critical condition, one serious, three fair, and two in good condition. A nursing supervisor said all suffered burns on their upper bodies, arms and hands.

Jan Feidler, who lives in a mobile home about 25 feet away, suffered minor injuries while trying to escape. Her home's vinyl siding was melted by heat from the flames. Ruiter was listed in good condition in Sacred Heart Hospital at Yankton, S.D. He refused to speak to reporters Saturday. Deputy state Fire Marshal Jim Holselaw was examining the remains of the storage tank Saturday. He said no cause had been determined but authorities presumed a spark ignited the gas. (Nebraska State Fire Marshal Report)

Liquid Nitrogen Asphyxiation, Springer, OK, September 1998

In early September 1998, an incident occurred involving liquid nitrogen, a cryogenic liquid that presents an asphyxiation hazard in addition to being very cold. The accident occurred while two workers were working on an oil pipeline in Springer, OK. They had been using liquid nitrogen in a pit to pressure check the pipeline and (in the opinion of the author) were likely to have been asphyxiated by the nitrogen displacing the oxygen in the pit. When discovered, both men were frozen by the very cold nitrogen, which has a liquid boiling point of 320 degrees below zero Fahrenheit; the vapors being released by the liquid would also have been very cold.

No final reports are available from any of the CSB's full investigations or incident reviews. Progress on these investigations can be followed on the CSB Web site (www.chemsafety.gov) or by calling the CSB office of external affairs at 202-261-7600. The mailing address for the CSB is U.S. Chemical Safety and Hazard Investigation Board, 2175 K. Street N.W., Suite 400, Washington, D.C. 20037-1809. Its fax number is 202-261-7650. While there certainly appears to be a legitimate function and identified need for the CSB, it is severely hampered in its investigative efforts because of limited resources. For the CSB staff members to become as effective as their NTSB counterparts, Congress and the Clinton Administration will have to provide adequate funding for this board. Investigations need to occur quickly after an accident to insure useful results. (Chemical Safety Board Investigation Report)

Bibliography

Volume 2

Atlanta Constitution, Georgia, April 8, 1894.

Burke, Robert, Burke Placard Hazard Chart.

Burke, Robert, CHEMTREC, Do You Have an Emergency Involving Chemicals? *Firehouse Magazine* October 31, 2001.

Burke, Robert, DOT Releases 2016 ERG, *Firehouse Magazine*, September 1, 2016.

Burke, Robert, Hazardous Materials Containers: Part 2 – Railroad Cars, *Firehouse Magazine*, March 31, 2000.

Burke, Robert, Hazardous Materials Containers: Part 3 – Fixed Facilities, *Firehouse Magazine*, July 1, 2000.

Burke, Robert, Hazardous Materials Containers: Part 4 – Portable Containers, *Firehouse Magazine*, September 1, 2000.

Burke, Robert, Hazmat Containers – Part 1: Highway Bulk Containers, *Firehouse Magazine*, January 1, 2000.

Burke, Robert, Hazmat Response Planning For Highways & Railways, *Firehouse Magazine*, September 1, 2014.

Burke, Robert, Intermodal Containers: What's Inside Those Big Boxes? *Firehouse Magazine*, November 30, 2004.

Burke, Robert, Kingman, AZ, Remembers 11 Fallen Firefighters, *Firehouse Magazine*, September 1, 2008.

Burke, Robert, The Southwest Boulevard Fire: Kansas City Remembers a Tragedy, *Firehouse Magazine*, November 30, 2009.

Chemical Safety Board (CSB), Herrig Brothers Farm Propane Tank Explosion, Albert City, IO, April 9, 1998.

Chemical Safety Board (CSB), https://www.csb.gov/.

Cranbury Press, New Jersey, December 30, 1887.

Cranbury Press, New Jersey, October 4, 1889.

Cranbury Press, New Jersey, August 29, 1890.

Cranbury Press, New Jersey, October 10, 1890.

Cranbury Press, New Jersey, May 22, 1891.

Cranbury Press, New Jersey, January 6, 1893.

Cranbury Press, New Jersey, January 12, 1894.

Daily Gazette Xenia, Ohio, February 24, 1897.

Deseret News, Utah, July 6, 1864.

DOT Emergency Response Guide Book, 2016.

Eau Claire Leader, Wisconsin, November 30, 1889.

EPA, Emergency Planning and Community Right to Know Act (EPCRA), https://www.epa.gov/epcra.

EPA, History: Federal Water Pollution Control Act Amendments of 1972, https://search.yahoo.com/yhs/search;_ylt=AwrCxGFiGC9eqkcAKC4PxQt.;_ylc=X1MDMjExNDcwMDU1OQRfcgMyBGZyA3locy1hdHQtYXR0XzAwM-QRncHJpZANZaWNMZWdwOFFVS1g2eGhKbThzdF9BBG5fcnNsdAMw wBG5fc3VnZwMxMARvcmlnaW5aW4Dc2VhcmNmNoLnlhaG9vLmNvbQRwb-3MDMARwcXN0cgMEcHFzdHJsAzAEcXN0cmwDNTQEcXVlcnlrkDRm-VkZXJhbCUyMFdhdGdGVyJTIwUG9sbHV0aW9uJTIwQ29udHJvbCUyME-FjdCUyMEFtZW5kbWVudHMlMjBvZiUyMDE5NzIEdF9zdG1wAzE 1ODAxNDQ5MjI-?p=Federal+Water+Pollution+Control+Act+Amendme nts+of+1972&fr2=sb-top&hspart=att&hsimp=yhs-att_001&type=att_pc_ homerun_portal_bucket_anim_nogl.

EPA, History: The Clean Air Act of 1970, https://archive.epa.gov/epa/aboutepa/ epa-history-clean-air-act-1970.html.

EPA, Laws & Regulations, Summary of the Comprehensive Environmental Response, Compensation, and Liability Act (Superfund), https://www.epa. gov/laws-regulations/summary-comprehensive-environmental-response-compensation-and-liability-act.

EPA, National Criminal Enforcement Response Team (NCERT), http://www. nerpca.org/NERPCA_Presentations/2016/kevin-gaul-epa-criminal-investi-gation-overview-2016.pdf.

EPA, National Oil and Hazardous Substances Pollution Contingency Plan (NCP), https://nepis.epa.gov/Exe/ZyNET.exe/100033RN.TXT?ZyActionD=ZyD ocument&Client=EPA&Index=2000+Thru+2005&Docs=&Query=&Time=& EndTime=&SearchMethod=1&TocRestrict=n&Toc=&TocEntry=&QField=& QFieldYear=&QFieldMonth=&QFieldDay=&IntQFieldOp=0&ExtQFieldOp =0&XmlQuery=&File=D%3A%5Czyfiles%5CIndex%20Data%5C00thru05% 5CTxt%5C00000000%5C100033RN.txt&User=ANONYMOUS&Password=a nonymous&SortMethod=h%7C-&MaximumDocuments=1&FuzzyDegree= 0&ImageQuality=r75g8/r75g8/x150y150g16/i425&Display=hpfr&DefSeek Page=x&SearchBack=ZyActionL&Back=ZyActionS&BackDesc=Results%20 page&MaximumPages=1&ZyEntry=1&SeekPage=x&ZyPURL.

EPA, Radiation Protection, Radiological Emergency Response, Radiological Emergency Response Team (RERT), https://www.epa.gov/radiation/ radiological-emergency-response.

EPA, Resource Conservation and Recovery Act (RCRA) Laws and Regulations, https://www.epa.gov/rcra.

EPA, Superfund, The Superfund Amendments and Reauthorization Act (SARA), https://www.epa.gov/superfund/superfund-amendments-and-reauthorization-act-sara.

Evening Gazette, Sterling Illinois, 1890-08–22.

Evening Star, Washington, DC, June 18, 1864.

FEMA, National Incident Management System (NIMS), https://www.fema.gov/ national-incident-management-system.

FEMA, National Response Framework, Fourth Edition, https://www.fema.gov/ media-library/assets/documents/117791.

Firehouse.com, https://www.firehouse.com/rescue/hazardous-materials/ article/12306150/lessons-learned-from-anhydrous-ammonia-incident.

Fort Wayne News, Fort Wayne, IN, May 23, 1895.

Fort Wayne News, Fort Wayne, IN, November 8, 1894.

Franklin, Ben A., Toxic Cloud Leaks at Carbide Plant in West Virginia, *New York Times*, August 12, 1985, Section A, Page 1, https://www.nytimes.com/1985/08/12/us/toxic-cloud-leaks-at-carbide-plant-in-west-virginia.html.

Fresno Bee, California, 1892-07–10.

GAO, U.S. Government Accountability Office, Hazardous Materials: 1990 Transportation Uniform Safety Act—Status of DOT Implementing Actions, https://www.gao.gov/products/145254.

Great Chicago Fire & Web Memory, October 8-10, 1871, Chicago, IL, https://www.greatchicagofire.org/great-chicago-fire/.

IMO, International Maritime Organization, http://www.imo.org/EN/Pages/Default.aspx.

Indianapolis Star, Indiana 1911-06–27.

International Campaign for Justice in Bhopal, Release of Methyl Isocyanate, Bhopal India December 2–3, 1984 https://www.bhopal.net/.

Iowa City Press, July 1875.

Jacoby, Anny, Marshall's Creek, PA, Nitrate Truck Explosion, June 26, 1964, http://www.gendisasters.com/pennsylvania/13388/marshall039s-creek-pa-nitrate-truck-explosion-june-1964.

Justice Department, U.S., USA PATRIOT Act: Preserving Life and Liberty, https://www.justice.gov/archive/ll/highlights.htm.

Legal Dictionary, https://legaldictionary.net/.

Marion Star, Ohio, February 26, 1889.

Marshfield Times, Wisconsin, December 18, 1912.

Maryland State Archives, Teaching American History in Maryland, The Great Baltimore Fire of February 8, 1904, http://teaching.msa.maryland.gov/000001/000000/000117/html/t117.html.

Mersereau, Dennis, July 10, 1926: The Day Nature Blew up a Town in New Jersey, The Vane, July 10, 2014, http://thevane.gawker.com/july-10-1926-the-day-nature-blew-up-a-town-in-new-jer-1602586498.

Middletown Daily Press, New York, April 19, 1892.

Montgomery Advertiser, Montgomery, AL, November 13 1912.

National Archives, Federal Register, Interstate Commerce Commission, https://www.federalregister.gov/agencies/interstate-commerce-commission.

National Fire Academy, Initial Response to Hazardous Materials Incidents: Basic Concepts, Student Manual, July 2003.

National Response Center, U.S. Coast Guard, http://nrc.uscg.mil/.

National Transportation Safety Board (NTSB), https://www.ntsb.gov/Pages/default.aspx.

New York State Department of Labor, Triangle Shirtwaist Fire, NY City, March 25, 1911, https://labor.ny.gov/agencyinfo/triangle-fire.shtm.

New York Times, New York, June 9, 1897.

New York Times, New York, August 4, 1855.

New York Times, New York, NY, July 1, 1890.

New York Times, New York, NY, March 22, 1892.

New York Times, New York, NY, July 3, 1886.

New York Times, New York, NY, May 4, 1895.

New York Times, New York, NY, July 5, 1864.

Newark Daily Advocate, Newark, OH, July 16, 1890.

Newark Daily Advocate, Ohio, September 8, 1897.

NFPA Standard 472, Standard for Competence of Responders to Hazardous Materials/Weapons of Mass Destruction Incidents, https://www.nfpa.org/codes-and-standards/all-codes-and-standards/list-of-codes-and-standards/detail?code=472.

NFPA, NFPA 704 Standard System for the Identification of the Hazards of Materials for Emergency Response, https://www.nfpa.org/codes-and-standards/all-codes-and-standards/list-of-codes-and-standards/detail?code=704.

NRT, U.S. National Response Team, Working Together to Protect against Threats to Our Land, Air and Water, https://www.nrt.org/.

OLA, Our Lady of the Angels Fire, Chicago, IL, December 1, 1958, http://www.olafire.com/home.asp.

Olean Times, New York, June 25, 1913.

OSHA Enforcement, http://www.osha.gov/enforcement. Code/Standard number and/or keyword search results.

OSHA Standard 1910.120, https://www.osha.gov/laws-regs/regulations/standardnumber/1910/1910.120.

Quincy Daily, Whig Illinois, October 8, 1890.

Quote Page One, Lacy, K., *The Road to High Reliability*, Distributed by Decomworld, p. 4, http://drillscience.com/DPS/KevinLacey-TheRoadToHighReliability.pdf (accessed May 30, 2018).

Republican Compiler, Gettysburg, Pennsylvania, November 30, 1846.

Rolla New Era, Missouri, December 22, 1888.

Sacramento Daily Record-Union, California, April 19, 1880.

United States Department of Transportation (DOT), PHMSA, https://www.phmsa.dot.gov/hazmat/erg/emergency-response-guidebook-erg.

Washington Post, District of Columbia, March 8, 1913.

Waterloo Daily Courier, Iowa, November 8, 1897.

Web Legal Advice, http://weblegaladvice.com/.

Webster's Dictionary, https://www.merriam-webster.com/.

West Virginia Historical Society, http://www.wvculture.org/history/wvhs/wvhssoc.html.

Index

after action analysis 184

after action follow-up 184

after action report 184

Agency Liaisons 134

Akron OH April 30, 1928 Benzene
    Explosion 281

Albert City IA April 8, 1999 Propane
    Explosion 77

Alton IL June 15, 1926 Refinery
    Explosion 279

Altoona KS February 12, 1924
    Nitroglycerine Explosion 276

Andale KS August 27, 1957 Farmers Coop
    Fire 308

Ardmore OK September 27, 1915 Gasoline
    Explosion 276

area command 121

Asheville, NC December 25, 1938
    Fireworks Plant Blast 289

Azusa CA August 21, 1946 Jet Plant
    Explosion 298

Baltimore City Fire February 7, 1904 18

Baltimore MD March 8, 1913 Harbor
    Dynamite Explosion 261

Baltimore, MD November 30, 1846 Powder
    Mill Explosion 244

Bayonne NJ December 29, 1960 Propane
    Explosion 312

Berlin NH January 23, 1934 Oxygen
    Cylinder Explosion 287

Berlin NY July 26, 1962 Propane Gas
    Explosion 314

Bhopal, India, December 2-3, 1984, release
    of methyl isocyanate 25

Big Heart OK January 26, 1919
    Nitroglycerine Explosion 270

Blakely GA January 27, 1970 Butane Gas
    Explosion 317

Blandford VA April 7, 1894 Fireworks
    Factory Explosion 252

Bristol TN March 1, 1947 Filling Station
    Explosion 299

Brooklyn NY May 31, 1970 Liquid Oxygen
    Truck Explosion 318

Browntown WI December 25, 1941 Silica
    Plant Explosion 292

Buffalo NY June 25, 1913 Grain Elevator
    Explosion 263

Burke Placard Hazard/Chemistry Chart
    24, 95

Camden NJ July 30, 1940 Camden Paint
    Factory Explosion 290

case studies
    Allegany County, PA Special
        Intervention Team 144
    Crete, NE February 19,
        1969 Derailment and
        Anhydrous Ammonia
        Release 171
    Miamisburg Ohio Phosphorus
        Incident 129
    West, Texas Ammonium Nitrate
        Explosion 151

Cedar Rapids IA May 1919 Douglas Starch
    Co Explosion 271

Clean Air Act (CAA) 1970 57
    summary 56

Clean Water Act (CSA) of 1970 and
    Amendments 34
    summary 34

Chemical Safety and Hazard Investigation
    Board (CSB) 73
    first investigations 74

CHEMTREC 87, 105

Chicago IL December 11, 1888 Oatmeal
    Mill Explosion 250

Chicago IL March 11, 1927 Chemical
    Company Explosion 281
Cincinnati OH July 5, 1864 Train
    Explosion 246
Cleveland OK March 9, 1916 Nitroglycerine
    Explosion 264
Clinton, IL February 21, 1925 Acid
    Explosion 277
Collinsville IL August 7, 1978 Propane
    Tank Car Explosion 325
command 115, 120
command staff 116
command post location 128
    Case Study Miamisburg Ohio
        Phosphorus Incident 129
commodity flow study 186
Comprehensive Environmental Response,
    Compensation, and Liability Act
    (CERCLA) of 1980 35
    superfund 35
    superfund, overview 35
Consequence Management 180
Coralville IA July 1875 Paper Mill
    Explosion 246
Crisis Management 180
Crossville IL January 14, 1952 Propane Gas
    Explosion 303
Cygnet OH September 8, 1897
    Nitroglycerin Explosion 258

Decatur IL July 19, 1974 Rail Yard Collision
    fire and Explosion 321
D.E.C.I.D.E. 161
decision making process 159
    data 159
        cognitive 159
        physical 159
        technical 159
Department of Energy (DOE) 72
Department of Transportation (DOT) 77
    chart 16, 90
    dangerous goods 86, 178
    hazard classes 88
    hazardous material 178
    hierarchy of hazards 89
    placards and labels 89, 90
Dispatcher/911 operator 24
    training 24
DOT Chart 16, 90
Drexel Chemical Company Fire & Explosion,
    July 5, 1979, Memphis, TN 111
Du Quoin IL February 15, 1941 Liquid
    Oxygen Plant Explosion 291

Dunbar PA February 22, 1966 Fireworks
    Plant Explosion 315
Dunreith IN January 8, 1968 Train Wreck
    and Explosions 316

East St. Louis IL January 23, 1972 Tank Car
    Explosion 319
Elliotsville WV November 30, 1889 Powder
    Explosion 253
Elizabeth NJ February 19, 1930 Oil Refinery
    Explosion 316
Emergency Operations Center (EOC) 180
Emergency Planning and Community
        Right to Know Act (EPCRA) 36
    40 CFR 65, 66
    fact sheet 37
    overview 37
    penalties 43
    requirements 44
    table 1: EPCRA chemicals and reporting
        thresholds 42
    title I 36
    title II 36
    title III 36
    title IV 26
Emergency Response Guide Book (ERG) 77
    evolution 78
Emergency Response Plan OSHA
    1910.120(p)(8)(ii) 193
essential elements of ICS 125
    accountability 128
    chain of command 126
    common terminology 125
    deployment 128
    integrated communications 128
    manageable span of control 127
    management by objectives 125
    modular organization 125
    Reliance on an IAP 126
    resource management 127
    transfer of command 128
    unified command 126
    unity of command 126
establish communications 151

federal regulations 65
    40 CFR-EPA 65
    29 CFR-OSHA 67
Federal On Scene Coordinator 150
Finance/Administration Section Chief 119
Fontana CA December 25, 1942
        Kaiser Steel Mills Butane
        Explosion 294

General Staff 116
GEDAPER 165
GEMBO 163
Great Chicago Fire October 10, 1871 17

Hartford CT May 21, 1892 Aetna
  Pyrotechnic Fire 255
hazardous materials containers 204
  fixed facilities 224
  highway 204
  portable 234
  rail 211
Hazmat Group Figure 2.52 126
  case study: Allegany County, PA
    Special Intervention Team 144
  hazmat EMS 150
  monitoring 149
Hazmat Incident Command 120
Hazmat Incident Commander 121
  command 120
  organizational structure 120, 126
  safety officer 123
  single command 120
  transfer of command 124
  unified command 121
Hazardous Materials Transportation Uniform
    Safety Act of 1990 (HMTUSA) 45
historical incidents 244
  incident history is a part of the
      standard of care 244
  1800s 244
    Baltimore, MD November 30, 1846
        Powder Mill Explosion 244
    Blandford VA April 7, 1894
        Fireworks Factory
        Explosion 252
    Blue Island, IL August 22, 1890,
        Standard Cartridge Plant
        Explosion 253
    Chicago IL December 11, 1888
        Oatmeal Mill Explosion 250
    Cincinnati OH July 5, 1864 Train
        Explosion 246
    Coralville IA July 1875 Paper Mill
        Explosion 246
    Cygnet OH September 8, 1897
        Nitroglycerin Explosion 258
    Elliotsville WV November 30, 1889
        Powder Explosion 253
    Hartford CT May 21, 1892 Aetna
        Pyrotechnic Fire 255
    Kings Station OH July 16, 1890
        Railroad Car Explosion 252

Louisville KY June 30, 1890
    Standard Oil Refinery
    Explosion 251
McCainsville NJ July 2, 1886 Atlantic
    Dynamite Co Explosion 247
Murray KY February 24, 1897
    Dynamite Explosion 257
Rochester NY December 30, 1887
    Naphtha Explosion 249
San Francisco CA May 22, 1895
    Nitroglycerine Factory
    Explosion 257
Tarrytown NY May 22, 1891
    Explosion and Train Wreck 254
Washington DC June 18, 1864
    Arsenal Explosion 245
West Berkeley CA July 9, 1892
    Powder Works Explosion 256
Wilmington DE August 3, 1855
    Powder Mill Explosion 245
1900s 258
  Akron OH April 30, 1928 Benzene
      Explosion 281
  Alton IL June 15, 1926 Refinery
      Explosion 279
  Altoona KS February 12, 1924
      Nitroglycerine Explosion 276
  Andale KS August 27, 1957 Farmers
      Coop Fire 308
  Ardmore OK September 27, 1915
      Gasoline Explosion 266
  Asheville, NC December 25, 1938
      Fireworks Plant Blast 289
  Azusa CA August 21, 1946 Jet Plant
      Explosion 298
  Baltimore MD March 8, 1913 Harbor
      Dynamite Explosion 261
  Bayonne NJ December 29, 1960
      Propane Explosion 312
  Berlin NH January 23, 1934 Oxygen
      Cylinder Explosion 287
  Berlin NY July 26, 1962 Propane Gas
      Explosion 314
  Big Heart OK January 26, 1919
      Nitroglycerine Explosion 270
  Blakely GA January 27, 1970 Butane
      Gas Explosion 317
  Bristol TN March 1, 1947 Filling
      Station Explosion 299
  Brooklyn NY May 31, 1970 Liquid
      Oxygen Truck Explosion 318
  Browntown WI December 25, 1941
      Silica Plant Explosion 292

historical incidents (*cont.*)
Buffalo NY June 25, 1913 Grain Elevator Explosion 263
Camden NJ July 30, 1940 Camden Paint Factory Explosion 290
Cedar Rapids IA May 1919 Douglas Starch Co Explosion 271
Chicago IL March 11, 1927 Chemical Company Explosion 281
Cleveland OK March 9, 1916 Nitroglycerine Explosion 264
Clinton, IL February 21, 1925 Acid Explosion 277
Collinsville IL August 7, 1978 Propane Tank Car Explosion 325
Crossville IL January 14, 1952 Propane Gas Explosion 303
Decatur IL July 19, 1974 Rail Yard Collision fire and Explosion 321
Du Quoin IL February 15, 1941 Liquid Oxygen Plant Explosion 291
Dunbar PA February 22, 1966 Fireworks Plant Explosion 315
Dunreith IN January 8, 1968 Train Wreck and Explosions 316
East St. Louis IL January 23, 1972 Tank Car Explosion 319
Elizabeth NJ February 19, 1930 Oil Refinery Explosion 283
firemen who made the ultimate sacrifice 308
Fontana CA December 25, 1942 Kaiser Steel Mills Butane Explosion 295
Greenville SC May 20, 1946 Propane Gas Explosion 297
Hollywood FL September 9, 1971 Dynamite Truck Explosion 318
Houston TX June 6, 1953 Fireworks Plant Explosion 303
Iowa City IA January 24, 1975 Propane Explosion 324
Jersey City August 17, 1915 NJ Oil Plant Explosion 266
Kilgore TX April 18, 1931 Oil Tank Explosions 285
Kokomo IN May 12, 1928 Explosion In Laundry 282
Langtry TX March 5, 1925 Quarry Explosion 277
Latham KS October 25, 1979 Propane Gas Explosion 326

Leonardo, NJ May 23, 1958 Nike Missile Explosion 309
Liquid Nitrogen Asphyxiation Springer, OK, September 1998 330
Long Beach CA March 1, 1931 Gas Explosion 285
Marcus Hook PA February 5, 1932 Tanker Explosion 285
Marianna FL August 14, 1953 Explosion And Fire Bottled Gas 305
Mitchell IN June 22, 1961 Lehigh Portland Cement Co. Explosion 313
New York NY July 18, 1922 Chemical Explosion 274
Norfolk VA January 8, 1931 Oil Barge Explosion 284
Norman OK June 5, 1934 Dynamite Explosion 288
Orwigsburg PA June 3, 1959 Propane Truck Explosion 310
Owensboro KY November 12, 1939 Glenmore Distillery Fire 289
Pensacola FL January 2, 1926 Newport Tar & Turpentine Co Explosion 278
Philadelphia PA October 10, 1954 Chemical Plant Explosion 307
Philadelphia PA September 8, 1917 Frankford Arsenal Explosions 265
Pittsburg PA August 20, 1924 Gasoline Explosion 276
Point Pleasant, WV December 22, 1953 Gasoline Barge Explosion 307
Portland OR June 27, 1911 Oil Company Fire 258
Raymond NE November 18, 1982 Propane Explosion 327
Rowesville SC May 25, 1983 Fireworks Truck Explosion 328
San Pedro CA February 15, 1927 High School Classroom Blast 280
Santee, NE August 28, 1983 Propane Gas Explosion 329
Sioux City IA April 1974 Grain Elevator Explosion 319
Sioux City IA December 15, 1949 Ammonia Plant Explosion 301

Smithfield NC March 7,
1942 Ammunition Truck
Explosion 294
South Charleston, W. Va., Nov. 6,
1941 Plant Explosion 291
Sparrows Point MD November 20,
1926 Oil Tanker Explosion 278
Speedway IN December 4, 1959 Air
Products Plant Explosion 311
Syracuse NY July 5, 1918 Split Rock
Explosion 267
Toppenish WA November 29,
1940 Dynamite Warehouse
Explosion 290
Turner ID January 19, 1927 Gas
Explosion In Mormon Hall 280
Vallejo CA November 8, 1918 Mare
Island Navy Yard Explosion 270
Versailles PA May 2, 1942 Torpedo
Plant Blast 292
Waltham MA March 7, 1948 Plastics
Plant Explosion 300
Waukegan IL November 27, 1912
Corn Products Plant Blast 260
Wellington KS June 7, 1922 Tank Car
Fire 274
Westport CT May 3, 1946 Trailer
Truck Explosion 296
Whiting IN July 10, 1921 Standard
Oil Explosion 273

ICS structure 121
branch 121
division 121
group 121
unit 121
incident commander responsibilities 115
incident generated plans 181
plan of Action 181
site Safety Plan 181
incident levels 168
incident management 157
why do we need Incident
Management? 157
Incident Management Team (IMT) 116
incident priorities 167
incident Stabilization 167
life Safety 167
protection of Property and
Environment 167
incidents that caused regulatory change 16
fires 16
Baltimore City Fire February 7, 1904 18

Great Chicago Fire October 10,
1871 17
Our Lady of the Angels Fire,
Chicago, IL Monday December
1, 1958 20
Triangle Shirtwaist Fire, NY City
March 25, 1911 19
hazardous materials 21
Bhopal, India, December 2-3, 1984,
release of methyl isocyanate 25
Institute, West Virginia, August
12, 1985, methyl isocyanate Gas
Release 29
Marshall's Creek, PA, June 26, 1964,
truck explosion 24
SW Boulevard Fire Kansas City
Kansas, August 8, 1959 22
Interstate Commerce Commission 30
Iowa City IA January 24, 1975 Propane
Explosion 324
isolation zones 169
access to zones 170
safe refuge areas 170
typical zone shapes 170

Jersey City August 17, 1915 NJ Oil Plant
Explosion 266
Joint Information Center (JIC) 151
Public Information Officer (PIO) 151

Kansas City Kansas, August 8, 1959 SW
Boulevard Fire 22
Kilgore TX April 18, 1931 Oil Tank
Explosions 285
Kings Station OH July 16, 1890 Railroad
Car Explosion 252
Kokomo IN May 12, 1928 Explosion In
Laundry 282

Langtry TX March 5, 1925 Quarry
Explosion 277
Latham KS October 25, 1979 Propane Gas
Explosion 326
laws 6
legal 10
pit falls 11
ramifications 10
legislative process 6
Leonardo, NJ May 23, 1958 Nike Missile
Explosion 309
liability 10
types 11
Liaison officer 117

Local Emergency Planning Committee
    (LEPC) 185
    community emergency response
        plan 185
    local emergency planning
        committees 185
Logistics Section Chief 119
Long Beach CA March 1, 1931 Gas
    Explosion 285
Louisville KY June 30, 1890 Standard Oil
    Refinery Explosion 251

malfeasance 14
Marcus Hook PA February 5, 1932 Tanker
    Explosion 285
Marianna FL August 14, 1953 Explosion
    And Fire Bottled Gas 305
Marshall's Creek, PA, June 26, 1964, truck
    explosion 24
Material Safety Data Sheets (MSDS) 104
McCainsville NJ July 2, 1886 Atlantic
    Dynamite Co Explosion 247
Military placard system 103
Mitchell IN June 22, 1961 Lehigh Portland
    Cement Co. Explosion 313
Morton Specialty Chemical Company
    April 8, 1998 Paterson, NJ 77
Murray KY February 24, 1897 Dynamite
    Explosion 257

National Fire Protection Association
    (NFPA) 7, 21, 94, 215
National Incident Management System
    (NIMS) 11, 113, 114, 120, 153
National Oil and Hazardous Substances
    Pollution Contingency Plan
    (NCP) 46
National Response Center (NRC) 48
National Response Framework 332
National Response Team (NRT) 49
    Chemical, Biological, Radiological,
        and Nuclear Consequence
        Management 55
    radiological emergency response team
        (RERT) 55
    regional response teams (RRT) 51
National Transportation Safety Board
    (NTSB) 73
Negligence 13
New York NY July 18, 1922 Chemical
    Explosion 274
NFPA 472 7, 9, 71
NFPA-473 72
NIMS 113

Norfolk VA January 8, 1931 Oil Barge
    Explosion 284
Norman OK June 5, 1934 Dynamite
    Explosion 288

Occupational Safety Health
    Administration (OSHA) 15
    hazardous substances 35, 36, 38, 39,
        44, 46, 50, 67, 185, 186, 196, 199,
        203, 209
    regulatory enforcement 15
Operations Section Chief 118
Orwigsburg PA June 3, 1959 Propane Truck
    Explosion 310
Our Lady of the Angels Fire, Chicago, IL
    Monday December 1, 1958 20
Owensboro KY November 12, 1939
    Glenmore Distillery Fire 289

Pensacola FL January 2, 1926 Newport Tar &
    Turpentine Co Explosion 278
Philadelphia PA September 8,
    1917 Frankford Arsenal
    Explosions 265
Philadelphia PA October 10, 1954 Chemical
    Plant Explosion 307
Picatinny Arsenal munitions explosion 265
Pittsburg PA August 4 Gasoline
    Explosion 276
Placard & Label system 90
Planning for hazardous materials
        incidents 111
    OSHA 1910.120(p)(8)(ii) 66
Planning Section Chief 118
Point Pleasant, WV December 22, 1953
    Gasoline Barge Explosion 307
Portland OR June 27, 1911 Oil Company
    Fire 258
Private sector resources and regulations 94
Public Information Officer 151
Public Protection Options 170
    evacuation 170
    shelter In Place 170

Quest Aerospace March 27, 1998 Yuma,
    AZ 76

Radioactive materials 72
Raymond NE November 18, 1982 Propane
    Explosion 327
Recognition Primed Decision Model 159
regulations 6, 30, 65, 94
regulatory 6
regulatory enforcement 15

requesting additional resources 150
Resource Conservation and Recovery Act (RCRA) 62
    laws and regulations 62
Rochester NY December 30, 1887 Naphtha Explosion 249
Rowesville SC May 25, 1983 Fireworks Truck Explosion 328

safety officer 123
    responsibilities of the Safety Officer 123
San Francisco CA May 22, 1895 Nitroglycerine Factory Explosion 257
San Pedro CA February 15, 1927 High School Classroom Blast 280
Santee, NE August 28, 1983 Propane Gas Explosion 329
SARA 36
scene control features 168
    access to Zones 170
    establish Perimeters 169
    isolation Zones 169
    perimeter Distances 169
    safe Refuge Area 170
shipping papers 104
Sierra Chemical Company Explosion Jan. 7, 1998 Reno, NV 76
Sioux City IA December 15, 1949 Ammonia Plant Explosion 301
Sioux City IA April 1974 Grain Elevator Explosion 319
site specific plans 193
    elements facility response plan 193
    Emergency Response Plan (ERP) 193
    OSHA 1910.120(p)(8) 193
    OSHA 1910.120 Emergency Response Plan 196
Smithfield NC March 7, 1942 Ammunition Truck Explosion 294
Sonat Exploration March 4, 1998 Pitkin, LA 76
South Charleston, W. Va., Nov. 6, 1941 Plant Explosion 291
Sovereign immunity 14
Sparrows Point MD November 20, 1926 Oil Tanker Explosion 278
Speedway IN December 4, 1959 Air Products Plant Explosion 278
Springer, OK, September 1998 Liquid Nitrogen Asphyxiation 330
standard(s) 70

NFPA 472 70
NFPA-473 71
NFPA-704 94, 96
standard of care 1
    definition 7, 30
    elements 11
    EPA 311 66
    historical incidents 244
    NFPA standards 7
    OSHA 1910.120 66
    regulatory basis 30
    SOPs/SOGs 11
State Emergency Response Commission (SERC) 37, 184
Superfund Amendments and Reauthorization Act (EPCRA) a.k.a. (SARA) 36
    EPCRA, fact sheet 37
    EPCRA, overview 37
Syracuse NY July 5, 1918 Split Rock Explosion 267

Tarrytown NY May 22, 1891 Explosion and Train Wreck 254
Terminating the incident 183
    critique 183
    debriefing 183
    recovery 183
    termination 183
Toppenish WA November 29, 1940 Dynamite Warehouse Explosion 290
Traditional Decision Making Models 161
    D.E.C.I.D.E. 161
    GEDAPER 165
    GEMBO 163
transfer of command 124
Triangle Shirtwaist Fire, NY City March 25, 1911 19
Turner ID January 19, 1927 Gas Explosion In Mormon Hall 280

UN/DOT Hazard Classification System 88
unified command 121
Union Carbide March 27, 1998 Hahnville, LA 77
U S Patriot Act 58

Vallejo CA November 8, 1918 Mare Island Navy Yard Explosion 270
Versailles PA May 2, 1942 Torpedo Plant Blast 292

Washington DC June 18, 1864 Arsenal
      Explosion 245
Waltham MA March 7, 1948 Plastics Plant
      Explosion 300
Water Pollution Control Act Amendments
      of 1972 33
   EPA established 33
Waukegan IL November 27, 1912 Corn
      Products Plant Blast 260

Wellington KS June 7, 1922 Tank Car Fire 274
West Berkeley CA July 9, 1892 Powder
      Works Explosion 256
Westport CT May 3, 1946 Trailer Truck
      Explosion 296
Whiting IN July 10, 1921 Standard Oil
      Explosion 273
Wilmington DE August 3, 1855 Powder
      Mill Explosion 245